Meet the Enchanting World of

CAVES AND CAVERNS

Dedicated lovingly to PJWJ and
Joe Cavers Everywhere

Caving

An Introductory Guide to Spelunking

by Donald Jacobson and
Lee Philip Stral

Harbor House Publishers
221 Water Street, Boyne City, Michigan 49712

Caving
© 1986. Harbor House Publishers
All rights reserved.
Published in the U.S.A.
ISBN 0-937360-08-2

Contents

Acknowledgments

We would like to take a moment to thank the people and organizations who helped to make this book something more than a mere manual.

For assistance in product and safety information our deepest appreciation goes to Jim Whittaker of Recreational Equipment, Inc., Seattle, the first American to climb Mt. Everest.

Also, we thank the management of Ohio Caverns, West Liberty, Ohio, and of Luray Caverns, Luray, Virginia, for their help in obtaining the photographs that grace the pages of this book. To the National Park Service, special thanks for the use of photos of Carlsbad Caverns, possibly the most incredible of all cave formations in the United States.

The National Speleological Society graciously extended its help; without its assistance in certain areas of research, this book would have been as dry as the deadest of dead caves.

A special word of thanks goes to Paul Dower of the Springfield, Massachusetts Explorers Club for assistance in retrieving information on a certain incident and for supplying us with data on fitness training.

Finally, we give special recognition to all Joe Cavers of the world to perhaps the most important people of all without whom we never could have completed this book.

Introduction

The development of caving as a sport followed man's natural curiosity about his environment and his desire to tackle nature on its own terms. It resembles other outdoor sports in many ways. It is as demanding on the body as mountain climbing, and equally exhilarating. It offers the amazing sense of discovery the skin diver feels while exploring the depths of the sea. Yet, while caving shares much with other sports, it is a totally unique adventure. Descending into the dark regions of the earth is quite unlike any other experience.

Why, then, has caving been relatively ignored in America's quest to pursue the outdoors? Quite simply, there is scant information available to entice new cavers to the sport. Successful caving requires more than a desire to go underground and, while there are more than 100 local clubs affiliated with the *National Speleological Society* (NSS), there has been no concerted effort to let the public know what it is missing or how it can become involved. Not that NSS or its clubs discourage new cavers. Quite the opposite is true. Experienced cavers welcome with open arms any novice who is truly interested in the sport. All the beginner has to do is seek them out.

This book will tell you what you need to know to enter the world and experience the wonders of caves. If you want to add *spelunking* to your list of outdoor sports, the pages that follow probably will tell you more than you need to know. If you want to go beyond the basics covered, numerous technical manuals explain, in great detail, the techniques and theories of caving. Many more discuss the science of caves. And, of course, there are the clubs, known to the cavers as *grottos*.

If you hesitate because you still believe caves are cold and damp places unfit for humanity, rest assured that this merely is a matter of perspective. While some caves are indeed cold, the average cave temperature is about 50° F., and this temperature holds throughout the year. Regardless of the weather outside, you can be relatively certain in assuming that the weather underground will be constant. About the only thing that will prevent most cavers from going underground is severe rain or snow storms above ground.

If you think there are no caves near you to explore, chances are that you are again mistaken. In virtually every state of the union, from New York to California, through Pennsylvania, into Illinois and up to the Pacific Northwest, there are caves—an estimated 50,000 total. Thus, the probabilities are that you are no more than a few hours from a good site, and each one has its own personality. Caves have been formed in limestone, marble, granite and sandstone. Some were created by water flow, others by lava flow. There are caves that nature carved slowly from ice and caves that were formed abruptly by earthquake. Each process creates a different size and form of cave, from large, rounded caverns to systems of long, twisting corridors. Within these formations an array of natural sculptures put even the greatest artist to shame. In their beauty and individuality, the shapes and colors are second to none.

Talking about caves and getting into one are two different things. You will need equipment with such exotic-sounding names as *carabiners* and *pitons*, as well as the more familiar ropes and ladders. No unusual clothing will be necessary, but you will have to select what you wear carefully. You also will need to learn how to walk, climb and crawl. These seemingly natural activities take on a new form when you enter a cave. Walking on flat land is different from walking on the rough terrain of a cave floor; using a rope to climb up a wall takes more effort than practicing on a flight of stairs; and the crawlways of a cave do not even closely resemble your old playpen.

This brings us to the most important aspect of caving: physical conditioning. All the equipment and technique available will be superfluous if your body is out of shape. You must be prepared to carry your equipment and your body for 12 hours or more at a stretch. You will be walking, climbing, crawling, stretching, twisting and turning. Whether you are chubby or gaunt, if your body is not conditioned for endurance underground, you may have difficulty getting out of the cave you got into.

Even considering the potential result of poor conditioning, there is a greater threat to the caver. Spelunking is not really as dangerous as it first seems. It certainly is no more hazardous than skiing, and perhaps much safer than mountain climbing. The caver's greatest danger is himself; stupidity is the number one killer in caves. Most accidents are the result of carelessness — the failure to check equipment before beginning the outing or its misuse once underground. A caver cannot expect this equipment to serve him beyond its capability, and knowing your equipment and exactly what it can and cannot do should be the first step in the caver's preparation. The same is true of your body. If you know your limitations, you will not dangerously exceed them.

Putting these easily avoided dangers aside, you still may collect your share of cuts and bruises. A basic knowledge of first aid techniques will meet most of your requirements. If the unthinkable — the serious accident — does occur, the one thing to remember is that there is very little you can do for the injured person while you are underground. Knowing how to handle an emergency and how and where to get help before you enter the cave will minimize the trauma of injury. All you do is keep your head and get the help.

It is these things—the basic knowledge, technique and common sense of caves and caving—that will be covered in this book. We'll show you how to take the caving experience back home with you on film instead of tearing apart Mother Nature's handiwork for your trophy room, where to find good caves and how to get involved with a caving group. After that, you are on your own, but once you have mastered the basics, there is no end to the excitement and enjoyment you can experience as a caver.

*   *   *

Have you ever stood on the lip of a dark passage leading down into the subterranean depths? When you were young, did you and your friends ever venture into the gloom—imagining that some sort of horrible creature or criminal lurked behind the next overhang waiting to grab you? How many times did you play Tom Sawyer to your next door neighbor's Becky Thatcher?

Most of you at one time or another have come in contact with a cave. The question is—How did you handle it?

Throughout our entire human history (and pre-history), we have been fascinated by caves and have developed innumerable myths surrounding them. In more recent times, caves have taken on another face with desperados using them as convenient and easily defensible hideouts from which to prey on the surrounding countryside.

Thus, it is natural for us to want to explore caves—to confirm for ourselves the myth, mystery and beauty of our spelean riches.

But, there are two ways to explore a cave (as in all things)—the right way and the wrong.

So many of us as children or novice cavers have had the following sort of experience that it is worth repeating here.

You and a buddy arrive at the mouth of X Caverns on a partly cloudy spring day. It is warm outside, so you are dressed in your normal togs —shorts, T-shirt, and sneakers (no socks). Your friend is decked out in similar fashion, except he is wearing suede desert boots. The very height of fashion—no?

Aside from your sartorial splendor, you have made provisions for your gastro-intestinal tract as well. Lunch is a few tuna salad sandwiches, some bananas and a cupcake, all carried in the ubiquitous brown bag. Your drinks? A few bottles of soda pop should suffice.

Being complete cavers, you have brought along a veritable plethora of extra gear, including a flashlight, a candle, book matches, 50 feet of clothesline rope, and a pocket knife. In addition, your partner has brought a huge ball of kite string to play out behind you as you walk along. And, anticipating some beautiful sights, you have brought along your little camera which is dangling off your wrist by its fake patent leather strap.

Perhaps out of fear, or maybe wisdom, you left a note at home telling your folks where you are and when you expect to return.

Obviously all-together, you and your cohort make the move. He leads off into the darkness, playing his flashlight along the floor to illuminate any pitfalls.

Suddenly, for you, everything goes a brilliant white as

your head slams into a low overhang. As the pain recedes, the throbbing begins and you have acquired your first caving momento —a goose egg.

The next step is to pick yourself up off the ground and muddle ahead with stars in your eyes. Sadly, your partner has gone on ahead and left you behind. Soon, you catch up to him at the top of a steep pitch. The incline is muddy and slippery. As you start down, your feet leave terra firma and beat you to the bottom by a scant tenth of a second. The situation is not bad, though. You are only in a small puddle of water, so you play a trick on the other guy and neglect to tell him about it. After he hits, you are once again a matched pair—soaking wet and muddy.

For the next 15 minutes, the two of you wander down a small passage obstructed by debris carried there by a now dried-up underground river. The branches grab at your legs and arms and rip your camera free. It drops with a thud, and as you pick it up, it sounds like a broken Thermos bottle—full of little rattles and clanks. So much for the best times of your life.

Your friend is calling to you—Have you seen the end of the kite string? Somewhere during the past 75 feet, the string broke, in the branches. Since you lack the flash light, you inform him that he can perform an anatomical impossibility with the remaining string. He ties the end to a tree branch and carries on.

Breaking out of the tangle, you come upon your friend sitting by the edge of a pool playing his light across the water. He is seeking a continuation of the passage on the other side of the pool. Finding it, the two of you intrepidly begin to inch around a ledge of rimstone. With a crunching noise, the rock gives way and the leader is in the drink!

You fish him out, laughing as you go. Continuing, you get to the crawlway that he spotted earlier. Scrunching down, you peer down the tube, finding that trying to discover the possibilities within is not unlike looking at your father's angry face through a soda straw.

You lead down the passage. About 20 feet in, you come to a low section that has both water and a thin layer of mud coating the bottom. As you dip through, you bump your head again. To add insult to injury, you realize that your trusty brown bag also has become a casualty when you feel a sickening softness and hear a squishing sound as you tentatively probe the spot where your lunch used to be. Pulling yourself the rest of the way through, you sit up and pull the bag free. Remnants of sandwiches and banana and cupcake tumble through the bottom of the now soaked bag.

No food, but you can still slake your thirst with a cool soda. Another disaster—no bottle opener. When it rains, it pours!

Your friend has, by now, left the tube and is pressing you to go on. As you wend your way along, questions begin to form in your mind. Is it safe to press on? Why have all the problems cropped up? When will we make it out? The answers are held only by the darkness.

Matters progress from bad to worse. As you step ahead, playing your flashlight along the wall, your foot sinks into some slimy muck. It keeps going down and down. The grip on your foot is somewhat akin to the hold of a member of the Fearsome Foursome on a fullback carrying the ball—a veritable death grip.

Up goes your knee. You wrench your foot free only to discover that you have left one thing behind—your sneaker! Then you make a major discovery. Cave mud is very reluctant to give up what it has taken. You simply cannot find the shoe no matter how hard you try.

Take a moment and look at yourself. You are tired, hungry, wet and muddy, minus one sneaker and plus two lumps on the head. Not a particularly pleasing sight. The other half of your team is no better. For the past 30 minutes, his suede shoes have lost all semblance of support and grip and are no better than snowshoes.

Inside the cavern, you begin to realize that the average temperature is only about 55 degrees. Chilling you to the bone, aided by the saturated condition of your clothing, the cave air is slowly sapping your strength and with it, your judgement.

Deciding whether to go on or not has become a major debate in your mind. The answer is obvious—quit while you are alive—but you are unable to think that clearly. The next few moments are crucial. Go on and get into real trouble or leave while you are physically able to do so.

Just around the bend is a major drop. Should you go over the edge and down into the depths? Your friend makes that decision. He ties the clothesline to a pillar and begins to clamber down the well, leaving you in the dark. You worry and listen for his call. Every so often, he hollers back that he is getting closer to a great passage and wants you to start downward. You say no.

Smart choice.

He has reached the end of his rope and still has 10 feet to go. Peering over the edge, you realize that the predicament is getting worse. The rope is beginning fray on the rock, and it is slimy and slippery. Your friend's best option is to try to make his way back up the rope.

Hand over hand, he pulls himself up the lifeline. Minute after minute, his agonizing groans drift up to your ears. Slowly, one hand appears over the edge. Soon another. You grasp them and drag him over the edge. He lays on the floor and pants for endless minutes. Periodically, he raises his head and looks around. He is beginning to hate the cave as much as you despise the dark surroundings.

Life for the two of you has reached its all-time low. At this moment, neither caver is very far from crossing that thin line. Still, you are in control.

You take the first step. Picking your buddy up, you turn around and start back. The obstacles are great, but the threat—this time a known quantity—is less. The mud, the pond, the gun barrel passage filled with branches, all pass by quickly.

Yet, there is one final hazard—the outside world. While you have been beating yourself to submission under ground, the sky above has been preparing a torrential deluge. As you start your exit, it opens up with a downpour that would make Thor proud.

The mud slide near the entrance disappears. In its place is a watercourse straight out of Disneyland. Water smashes into you like a pile driver. Slowly you force your way up the slide and into the main entrance. Already, the level is rising. A river instead of a dry walk faces you. With your friend in tow, you make your way out into the world of light and people.

All that remains of your day in the cave is pain and the memory of near disaster.

All that you want is a warm meal and a dry bed.

All that you get is rain in your face.

But, you have learned. Never challenge a cave except on its own level. Meet it on its ugliest terms. Prepare yourself for everything that only a cave, in its own perverse way, can throw at you. Steel yourself for rigors that you never will experience anywhere else.

Read the rest of this book and you stand a good chance of winning when you go back.

Chapter 1

The World of Caves

Caves, like human beings, come in all sizes, shapes and colors. You'll find caves in a variety of surroundings, often in places you would least expect one to be lurking. Perhaps it's this uncertainty that adds spice to prospecting for caves. They appear in odd forms carved out of limestone, marble, sandstone, lava flows, ice and even snow. They can be found among the peaks of great mountains, beneath the plains, on the seashore and under water. The combinations are endless. Of interest to most cavers, though, are the caves found in limestone. These will be discussed in depth in this chapter, along with the rarer caves found in lava beds, ice and the bluffs bordering on the sea.

Before going any farther, there is one type of cave that you should never consider exploring unless you are 1) an excellent scuba diver, or 2) slightly crazy. These are the *sea caves,* caverns found under water. There are many such caverns off the coast of Florida, and they have claimed many lives as the price for their exploration. While most land caves show little mercy for mistakes, they at least allow some margin for error. A sea cave shows no mercy for the ill-prepared or foolhardy. It will kill you. Don't go into one and shun anybody who suggests such an expedition.

Where to Find Caves

Caverns of some form or another have been discovered in all 50 states. Some are merely tiny holes or crawlways, while others are great networks of immense rooms connected by long corridors. The prime caving regions of the United States can be broken into two main groups (see map, Major Caving Areas of the U.S.A.). The first and foremost area forms a crescent that begins in the Blue Ridge Mountains and Shenandoah Valley of Virginia and

Major caving areas of the United States.

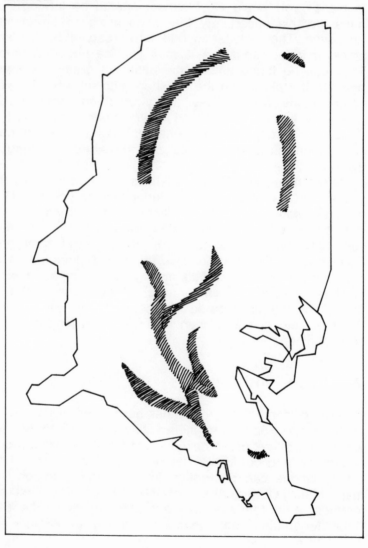

West Virginia, dropping down into the Great Smoky Mountains of Kentucky and Tennessee. This rich caving territory then takes a sharp turn west, spreading across the two states until it reaches the Mississippi River. Along the way, the crescent is scattered with hundreds of caves in southern Ohio, Indiana, northern Georgia and northern Alabama. Finally, the region peters out after crossing the Mississippi River into Missouri (if you can call more than 1,000 caves "petering out"). All told, more than 7,000 caves have been discovered in this, the most fertile of caving crescents.

The other major caving region is found in the Southwest. Again, this area forms a crescent from the Gulf of Mexico on the Texas coast and through New Mexico, Colorado, Utah, Arizona and Nevada, to California. About 1,600 caves are found in this region.

If you do not live near these areas, do not despair; no area is immune from the forces of nature, and your state is sure to contain many caves. Even tiny Connecticut has more than 10 reported caves, with New York's large limestone caverns only a short drive away. Northern Illinois and southern Wisconsin sport many fine caverns in their Mississippi River areas, and Washington and Oregon boast extensive lava tube systems in the Mt. Rainier and Mt. Hood areas. Regardless of where you are, a cave of some sort probably can be found nearby. More information on locating caves in your area is available by writing to the National Speleological Society or one of its local chapters (see chapter 8).

The Giants: Limestone Caves

If you were to run an all-time size and scope contest among caves, the great limestone systems like Mammoth-Flint Ridge, Carlsbad and Luray certainly would be in contention. It just seems to be the nature of the beast that *limestone caverns,* given the right conditions, grow to immense proportions. Many give no immediate clue to their actual size, having rather modest openings and entrance passages. They branch out without warning and just seem to keep on going. Mammoth-Flint Ridge is just one of these caves (actually a system of interconnected caves) that can be found by enterprising spelunkers. It has been actively explored for almost 200 years, and still shows no sign of ending.

Limestone is found throughout most of the world. It is the result of countless eons of organic residue that collected on the floors of the great seas that once covered most of the earth. As the ages passed, the lower layers of residue decayed and were compressed by the mass above into a hard, flinty stone that is primarily *calcium carbonate.*

Once the ancient seas dried up, the limestone was covered by a variety of rocks and minerals. Gradually, the outer layers eroded and the prime ingredient for cave formation, water, was introduced. The water entered the limestone through a number of avenues, usually along cracks or *faults* in the rock itself. As the water came in contact with the rock, some of the limestone dissolved, forming *carbonic acid.* This, in turn, dissolved more of the limestone mass as the water and acid percolated downward in their continuing search for the lowest possible point. This process of *dissolution* and development of underground drainage is called the *karst process,* the term coming from a plateau on the Adriatic coast of Yugoslavia that exhibits this process.

Karst lands are easy to spot, and they provide the enterprising cave detective many clues as to the whereabouts of caverns and underground rivers. The plateau of central and southern Indiana is a prime example. It is marked with pits, *sink-holes,* pirated streams and other characteristics common to karst lands. Generally, the surface of the area is flat, the only varying features being the small rounded knobs indicative of karst topography. The underground drainage of this area is immense, with streams disappearing from and returning to the surface almost at will.

Dissolution, however, is not the only action that expands a cave. The underground streams rush through deep passageways, bearing down on the rock around them until broad avenues are formed. The stream then either dries up or is lured away by another drainage system, leaving a cave for us to explore. This solution process of

Process of Dissolution: A: Simple fault (crack) line in the bedrock. B: Dissolution begins, but is stopped by a layer of nonporous granite. C: As the limestone bedding continues to deteriorate, sinkholes appear on the surface as the bedrock collapses. D: Finally, as the cave expands, water breaks through the granite to continue the process in the lower layer of limestone. Although the lower half of Stage D is still water filled, it will develop into an extensive cave as dissolution continues and the water level drops.

Limestone cave passage.

scouring and dissolving takes millions of years and, many caverns later, it still continues. Thus, when you hear some grimy character speak of a *living cave,* he's talking about one in which water is still active, as opposed to a dead cave, like Carlsbad, which is as dry as the desert above. Happily, there is plenty of limestone on this earth, and most of it is being dissolved for approving cavers.

This dissolution process also is responsible for other attractions caves hold. Without it, there would be no sights worth seeing; spelunkers would be reduced to clambering around bare walls. The ever-present cave flora is created by the slow, but persistent, solution and precipitation of calcium carbonate—not life made out of rock, but "living" rock. Such beauties as *stalactites* and *stalagmites, helictites* and *culupholites* are all forms of cave flowers called *speleothems.* Actually, any of the delicate or imposing shapes formed by dripping water laden with dissolved limestone as a result of solution can be classified as a speleothems.

Draperies and curved draperies.

Frustrated Column, King's Palace, Carlsbad Caverns National Park. (photo courtesy of Carlsbad Caverns National Park)

Big Room - Onyx Draperies, Hall of Giants, Carlsbad Caverns National Park.
(photo courtesy of Carlsbad Caverns National Park)

*The fabulous stalagmite Rock of Ages
at Carlsbad Caverns was first seen by
an intrepid cowboy named Jim White
almost 100 years ago.
(photo courtesy of the National Park
Service)*

*The Crystal King stalactite in Ohio
Caverns. This is a prime example of
the pure white gypsum stalactites
unique to Ohio Caverns.
(photo courtesy of Ohio Caverns)*

Of all the speleothem's varieties, stalactites and stalagmites are the most familiar and the most common. The ici-
cle-like formations hanging from the ceiling or thrusting
from the floor of a cave can be as tiny as a pin or as
massive as the great mounds in Carlsbad. They are
formed as water seeps out of cracks and evaporates, leaving behind a deposit of *calcite* (the mineral form of calcium carbonate). Most other minerals in solution, including
sodium and magnesium, also will form a stalactite or stalagmite. Stalagmites are formed as water drips to the floor
and evaporates, leaving a cone-shaped pile of the mineral.
Helictites and culupholites are similar formations, but
these types grow in undisciplined disarray. Their chaos is
beautiful, but extremely fragile. *Gypsum needles,* wormlike
growths and other protruding deposits are all members of
this group.

Between stalactites and helictites are other formations,
such as *draperies* and *shields,* so-called because the deposits of stone are surprisingly similar to curtains of cloth
hanging from the ceiling. At times, they form impenetrable
barriers across promising passageways, forcing the caver
to seek another way around the wall.

Helictites.

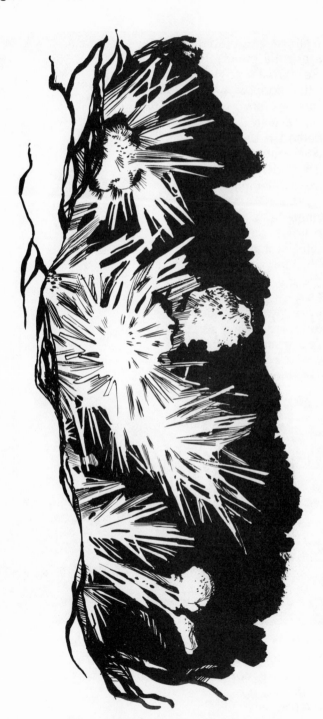

Aragonite crystals.

Stalactites at Cave of the Mounds growing toward the floor will eventually become columns. (photo courtesy of Alan Bagg.)

*"Soda Straw" stalactites.
(photos by Donald Jacobson)*

Many other cave formations abound within the confines of this subterranean world. *Speleogens* are formed by the wearing and tearing action of wind and water. They are acute physical characteristics of the caves themselves, appearing as ripples caused by flowing water or potholes caused by pebbles thrown into the cave walls by swirling water. A third type of formation, is the *petromorph*. Petromorphs are secondary geological characteristics normally found in less profusion than speleogens. A common type of petromorph is the *boxwork* found near cave entrances and created when sections of soft bedrock are eroded by wind and water with the remaining ridges of hard bedrock taking the form of open boxes. Finally, there are barriers caused by rock falls and clay deposits. These also are known as *clastic fills* and can be a major headache to a caver bent on breaking a new trail. Invariably, clastic fills crop up in the midst of a tiny crawlway, blocking a promising passage with several feet of rubble. Not only is it inconvenient to crawl backwards, it also is quite frustrating to see a broad passage on the other side of the fill, with no way to reach it.

Speleothems.

One other type of formation falls under the speleothem category, but it is a constant source of semantic arguments. This is the *column.* A column is a stalactite and a stalagmite fused together to form a solid, vertical piece of rock. The controversy arises because many people confuse them with pillars, or vertical pieces of bedrock. Pillar or column, only the most dedicated speleologist will take you to task for mistaken identification. The point is that all these features combine to give limestone caverns the beauty and life that make the exercise in rock-climbing technique worth the effort. And there are yet other cave types to give even the most dedicated caver a varied fare each time he enters the world beneath the surface.

Littoral Caves

Littoral caves are often misnamed "sea caves." Perhaps they should be called "caves by the sea," since they are found along every seashore in the world, offering the less ambitious caver an opportunity to explore a unique type of cavern with minimal fuss and bother. Some of the more famous littoral caves are the Blue Grotto of Capri and Fingal's Cave of Scotland. In the United States, excellent littoral caves are found on both coasts from Anemone Cave in Maine to Sea Lion Cave in Oregon. California seems blessed with an abundance of these caves, especially in the coastal cliffs near La Jolla and on the islands in the Santa Barbara Channel. Other, less accessible littorals can be found along the Alaskan coast and on the Aleutian Islands.

Entrance to a littoral cave at La Jolla, California.
(photo courtesy of Donald Jacobson)

Not all littoral caves are found by the shining sea, however; for example, Clinton's Cave was formed when the Great Salt Lake was fresh water and several hundred feet deeper. Yet, this is a true littoral cave and bears all the signs of a cave still being formed as well.

Geologists say that littoral caves are formed by one of three processes—hydraulic action, *corrasion* or corrosion. Of the three, hydraulic action is the most prevalent and forceful. As the waves break on the base of coastal cliffs, water is driven forcibly into every opening with incredible power. (Waves in the Atlantic exert some 2,000 pounds of pressure per square inch in calm weather and more than 6,000 pounds during a storm.) This action wears holes and passageways into the rock. The water always follows the line of least resistance, usually through cracks and faults in the rock itself, eventually creating *gloups* or throats. These are more commonly known as *blowholes,* with one opening on the surface behind the cliff and the other on the face of the cliff. In time, sections of the gloups collapse, forming a cave, or the water wears away enough rock to connect with another throat, forming a natural bridge. The danger of high surf is always present in these caves but, in some cases, the characteristics of the area around the cave may change so that it is shielded from the rampages of the sea.

Corrasion and corrosion play a lesser role in the creation of littoral caverns. Often, the rock is granite, not limestone, and therefore is less susceptible to dissolution by acid. And, while the sea is fierce, it must truly be enraged to pick up boulders and hurl them at the cliff for the process of corrasion to take place.

Three words of warning about exploring littoral caves: *Remember the tides.* The fact that you are inside exploring will not change the natural pattern of the earth. You could be trapped in a littoral cave that is filling rapidly with water as the tide rushes in. Don't get caught; find out the exact times of high tides on the day you plan to enter the cave. It could save the rescue people a lot of time and trouble.

Lava Tubes

An intriguing form of cave that is just beginning to gain recognition by spelunkers is the *lava tube* cavern. These wonders can be as small as a single man or as large as the bigger limestone caverns. Some of the tubes are not tubes in the normal sense, but avenues 100 feet wide and 30 to 40 feet high. Obviously, they can be found only in areas that have experienced volcanic activity, which in the United States is generally limited to the West Coast and Hawaii. In these areas you will find extensive tube systems, especially in the Lava Beds National Monument in California and the lava flows near Mt. Hood, Oregon, and Mt. Ranier, Washington. The king of them all, Kazumura Cave in Hawaii, is more than two miles long. Favorites among (get ready, now) vulcanospeleologists are Ape Cave and Dead Horse Cave in Washington.

A variety of sights can be seen in lava tubes that are never to be experienced in limestone caverns. In lava tubes, normal formations like ripples and splash marks were not formed over the ages by slowly dripping water, but in minutes and hours by rapidly cooling magma. Frozen lava falls in the tubes look like water, but never move. Meatballs hang suspended between outcroppings where they were left by the lava flow thousands of years ago.

Lava tube caverns are formed within hours of the initial cooling of a fresh lava flow. As the river of molten rock begins to cool, the outside hardens first, and the liquid core of the mass begins to flow out, not unlike water draining from a pipe. Eventually, the interior drains away completely, leaving a tube behind. Lava tubes also can be formed by geothermal activity, with gas being forced through cooling flows, leaving tubes that may remain active as *geysers, mudpots* or other volcanic creatures. Explorers in Hawaii, in an effort to learn more about the formation of tubes, have ventured into them after they have cooled just enough to prevent incineration of the human body, only to be thwarted by sulfurous gas. On the whole, though, lava tube caves offer a delightful, albeit murky, alternative to normal caving fare, and since the pitch black lava is much harder than limestone on equipment and clothing, as well as on bare skin, exercise more than usual caution should you decide to explore these ideas.

Ice Caves, Etc.

Ice caverns are perhaps the most fascinating and fragile of any cave type yet discovered. Found in glaciers and high mountain snow fields, they are embellished with crystalline structures not found anywhere else, and they almost come and go with the seasons. Usually, ice caves are formed by the action of streams flowing under the ice, but swirling mountain winds also can form them. Summer air, passing through exposed corridors, contributes to the growth of the passageways.

Ice caves can be found in most high-county regions that both receive enough snowfall and remain cool enough year around to maintain the snow and ice. The Paradise Ice Caves in the Paradise Glacier on Mt. Rainier are perhaps the best example of ice caverns in the country.

Talus caves or caves formed by the collection of fallen rock, really are not caves in the normal sense of the word. They never experience absolute darkness, although they can be pretty black. Several such cave systems have been commercialized, most notably Plymouth Caves in New Hampshire, and these are relatively safe. Extreme care must be exercised, however, in exploring noncommercial talus caves, as the rock may be unstable and slippage could occur.

Last, but not necessarily least, *block creep caves* are formed when two sections of rock separate along a fault line (as in an earthquake), leaving an opening of varying width and depth between the two units. This type of cave is best suited for vertical caving, but there are many horizontal examples to be found. Generally, these slots offer little challenge, but the Adirondack Mountains of New York hold several promising examples.

Flora and Fauna

Life in caves can be delineated much the way a cave is divided—mouth (daylight), entrance passage (twilight) and interior (eternal darkness). Plant life, because it needs light and cannot survive in the absolute darkness of the main body of a cave, is confined to the entrance areas and is not unique in any sense. Cave fauna, on the other hand, is diverse and unusual. Insects, amphibians, mammals, reptiles and fish all can and do live in the deepest of American caves. They have had to adapt to unusual conditions and often they only vaguely resemble their brothers and sisters on the outside. And while most cave life is no threat to man, there are creatures who inhabit certain sections of caves and can cause some uncomfortable moments.

At the cave entrance, you will find most forms of wildlife indigenous to the area, be they birds, bears or snakes. Generally, they enter the cave mouth to seek shelter from the elements or to hide from diligent hunters. In the southwest, you would be well advised to watch closely where you step as you enter a cave, to avoid disturbing one of the regular inhabitants e.g., your friendly neighborhood rattlesnake.(Hint: If it's a sunny day and you must climb to the entrance, carry a long stick and sweep the rocks above your head as you climb. That way, you won't surprise a snoozin' snake.) The residents of the cave entrance are transient by nature, rarely making the cave a permanent home. However, some bears do make dens in caves, and it is wise to check with local residents before you try one on for size. Other critters won't bother you if you don't corner them. Let them leave, then go caving.

In the twilight zone, you can find some of the more permanent cave residents. Usually, they are the nocturnal creatures—mice, rats and insects. These little fellows are quite happy where they are and would appreciate it if you did not wipe out their burrows and nests through indiscriminant activity. They are, however, as equally at home on the outside as on the inside and have not undergone any physical changes to adapt to the environment. They have not made the final commitment to total darkness and, on a busy night, they will be in and out as they sleep and eat.

Bats, while often the material of horror, also are the symbol of cave life to millions of people and, as long-time twilight zone residents, the creatures deserve some special attention. Only one species of bat—the Vampire found

Entrance to Carlsbad Caverns. (photo courtesy of Carlsbad Caverns National Park)

in Central America—drinks blood, but the primary danger of bats is that they are major carriers of rabies, and for this reason, a bat bite should be regarded as a medical emergency. Fortunately, bats flee from human intruders, and bat attacks are rare. If you care to study them, hundreds of different species are found in caves.

Bats serve a vital function in the food cycle. Farms near caves experience less crop damage from insects thanks to the bats that come out at night and feed on the bugs. Bats also have been instrumental in revealing caves' mysteries to speleologists. From the pattern of bat colonization, scientists can determine the geological changes that have occurred in the cave. And bats have helped us discover some of our largest caverns. Carlsbad was discovered by a cowboy intrigued by a funnel-shaped black cloud rising out of the earth at dusk every night. The cloud actually was millions of bats exiting to forage for food. A final note on the usefulness of bats to man is their guano. The stuff stinks, but guano is high in nitrates used to manufacture explosives. Before going commercial, Carlsbad was mined for bat guano.

While bats are not necessarily harmful to you, you can be harmful to them; so never disturb a colony while it is hibernating. They will not harm you, but you could kill them by waking them up. Awake, bats burn up energy— energy that can be replaced only by feeding, and the insects that make up a bat's diet just aren't around in the winter.

As we descend farther into the deep, a different set of life forms appears—the animals that live their entire existence without ever seeing the light of day. Over the eons, they have adapted to the peculiar conditions of lightless living, and they never leave the cave out of choice, staying inside to seek food. Salamanders, fish and crustaceans are the sole residents deep within the cave, except for plankton-like beasties found in some caverns. Most often, they have no developed eyes and little or no body color. Instead, these creatures have developed supersensitive sensory systems attuned to vibrations in their environment.

Many of the cave creatures thrive on waterborne food such as waste and other dead creatures. Many cave dwellers are microscopic in size and thus beyond our scope

here, but you will find fish of varying sizes swimming in
the pools deep in the caves. Crayfish cruise the floors of
the puddles, and shrimp populate the waters themselves.
When caving, you must remember that this is a very fragile ecosystem—one thing out of balance and the whole
cave will die. That's why you *never* take any cave life from
the cavern or leave any waste that might endanger its
population. To preserve the beauty of all-white or trans-
parent creatures swimming about, be careful with all your
equipment.

While we have been discussing animal life in caves, one
animal has been noticeably missing—man. Not too many
men live in caves nowadays, but in prehistoric times,
caves provided adequate and extremely low-cost housing.
Most of the cave dwellings discovered to date are in Europe, but there are notable cave dwellings in Arizona and
New Mexico. These are the *pueblos* or stone houses, of
Indians who vanished from the region centuries ago. The
homes are constructed from stone and adobe and are set
in the entrances of massive overhang caves found in the
cliffs of the now-desert region.

Nevertheless, Europe, and to some extent Asia, remain
the main areas for cave dwellings and they have stories to
tell. Archaeologists have determined that the first attempts
at cave habitation were simple shelters for the beleag-
uered men and women, away from marauding animals and
the ravages of the weather. Later, as man learned to use
fire and build tools, the caves also served as workshops.
Finally, within the last 15,000 years, as man began to develop a rudimentary civilization, caves became the center
for religious activity. The great cave paintings of Lascaux
in France have been attributed to the desire for a good
hunt. Other French caverns have housed statuettes of animals and people, with still others revealing the grisly reali-
ties of prehistoric life—human sacrifice and ritual murder.

In America, caves were used by the Indians as a ready
source of flint for arrow and spear heads more often than
for shelter. Evidence of prehistoric mining operations in
caves in Indiana (especially Wyandotte Cave) include pre-
served footprints and the mummified remains of a flint
miner trapped by a rockfall. More recently, with the arrival

of the white man, caves have assumed another role, usually a violent one. Mammoth Cave in Kentucky, a beautiful cavern, and perhaps the most extensive one in the world, was mined for decades in the early 19th century to strip it of its nitrate wealth. Historians may say the saltpeter excavated from Mammoth Cave prevented an American defeat in the War of 1812, when a blockade cut off our foreign supply, but one cringes at the thought of the countless formations destroyed by man as he looted the cave of its mineral wealth. Other caverns were mined similarly for this mineral vital in the manufacture of gun powder. Still others served as hideouts for outlaws and their loot. Jesse James seems to have had a peculiar affinity for caves, and there are several bearing his name in Missouri and Kentucky (not unlike "George Washington slept here"). The most notorious cave of all is Illinois' Cave in Rock located in the bluffs above the Ohio River near the Indiana border. As early as the 1790's, outlaws inhabited the cave while preying on the river flatboat traffic. They would entice weary boat crews with promises of liquor and women, then murder them and make off with the cargo. Later, the infamous brothers Harpe, who indiscriminantly murdered and plundered, made the cave their headquarters. With the advent of steamboats and railroads, however, the cave's nefarious utility disappeared and, with it, the outlaws. The cave is now a state park and commands one of the most beautiful views of the river and the surrounding region.

Today, caves are famous for other reasons—their depth, distance and difficulty. Depths change constantly but, at press time, the *Berger* in France held the record for the deepest chasm from entrance to bottom; explorers have pushed more than 3,000 feet below the entrance to this cave. Another depth record—longest free drop—now held in the western hemisphere, also has been changing. *El Sotano* in Mexico currently holds the record with a 1,345-foot drop (for comparison, visualize a 135-story building). As for distance, several caverns could be number one — Mammoth-Flint Ridge, Holloch in Switzerland, the Greenbriar System in West Virginia or South Dakota's Wind Cave. Many still are being explored and are growing by leaps and bounds.

Difficulty is a relative concept, but most cavers agree that West Virginia's Schoolhouse Cave probably is the most taxing. Sweet Schoolhouse is a single horizontal shaft with innumerable pits and crevasses. The only way to move forward is to crawl, descend, ascend or traverse, in whatever combination that fits. The physical effort required from a caver here pushes many beyond their endurance. Fortunately, most caves in America fall well below these bounds.

Chapter 2

What You'll Need

If you've read this far, you probably have some desire to descend into the bowels of the earth. But before you get into a cave, you will need certain equipment to ensure your safe entry and return. And making the trip comfortable is not a bad idea, either.

Caving combines the elements of a number of sports—mountain climbing, hiking, swimming and rock climbing, to name a few. Thankfully, you do not need everything each sport requires, but without organization and a knowledge of what you *do* need, you might end up looking more like a safari porter than the enterprising explorer you really are.

Covering Your Body

First, look at yourself—head to toe. That sporty denim outfit looks great, right? The epitome of fashion for the Great Outdoors. Take that into a cave, however, and it will be torn to shreds after five minutes of crawling, not to mention what your head would resemble after taking a few bumps. Obviously, you need some sturdy clothing and headgear.

A bit of basic cave philosophy requires that you think warm. Selecting apparel for a caving expedition requires a similar state of mind as in planning for a winter camping trip. This, in turn, indicates either a one-piece union suit or the traditional long johns with matching shirt. This could be the most important part of your outfit. Without it, you literally could freeze to death in a cave and never realize it. The pictures you see of tourists cruising around caverns in shorts and t-shirts are valid only in that they were shot in a cave. The average tour of a commercial cave is about one-and-a-half hours, not enough time to develop a sound chill, especially when commerical caves are heated by all sorts of devices.

A standard cave exploration, however, often is six to eight hours of slogging through mud, sometimes up to your neck. The standard condition of human life in a cavern ranges from soggy to downright saturated. Combine that with an average temperature of 50° F. and a draft, and it can become uncomfortable in a cave if you are not properly prepared. Actually, the word to use is not "uncomfortable." The word is *hypothermia*. Hypothermia is discussed in depth in chapter 6 but, quickly, it is the rapid loss of body heat, to the point where body temperature drops to below the critical level and the victim freezes to death.

One way to avoid hypothermia is to wear effective insulating clothing, of which thermal underwear is only one component. Another is at least one, if not two, wool or flannel shirts. Again, these will insulate you and prevent excessive loss of body heat. As you begin to work up a sweat with the strenuous activity of exploring the cave, you can always strip off a shirt or two, and you can put them back on as you begin to feel cool. But if you didn't wear them in the first place, how could you even approach warmth?

Some people suggest that a pair of jeans should be added before you pull on your outer wear. These are optional, but the next part of your uniform—coveralls—is not. There are two schools of thought on the use of coveralls. Many European cavers eschew them and, for that matter, any other clothing. This is because many caves in Europe are simply tiny crawlways and *catwalks* liberally covered with slime and muck. Clothing has a tendency to hinder rather than help in such situations and perhaps skin slides along mud more easily than denim. But there is no way you will want to expose your skin to the grinding and grating effects of good ol' American limestone. Wear coveralls—and *not* the polyester wonders of the fashion world currently so popular. Caving coveralls should be made of sturdy, double-weight denim cloth. They should be closed by a half-length zipper rather than buttons since zippers keep mud out more thoroughly.

Most standard coveralls come with pockets and flaps, and you should cut off the flaps and sew up the pockets. First you won't be carrying anything in them (cave walls love to catch lumps and halt progress) and, second, pockets and flaps have the nasty habit of catching on protrusions which either halts progress or ruins the clothing, exposing your skin to the harshness of nature.

Much of this information comes from hard experience. Until you have discovered the subtle pleasures of struggling for half an hour to free yourself from an overhang caught on your pants pocket, you may not understand. Not only have we been hung up while negotiating an excrutiatingly narrow passage, but caving buddies bringing up the rear (so to speak) had the dubious pleasure of viewing an exposed posterior after the cave won the battle and had taken its toll on the bottom half of the coveralls. Naked flesh gets cold very rapidly.

After the body come the feet, perhaps the most tender part of your anatomy. As most hikers know, blisters adhere strongly to the theory of spontaneous generation. They seem to spring up out of nowhere at the most inopportune moments, making it painful—sometimes near impossible—to walk. Blisters are caused by two well known agents—loose shoes and wet feet. There is no cure in most caves for the latter, but you can compensate for the former.

Many people try to go caving wearing sneakers as their primary footgear. Admittedly, sneakers are light and comfortable, but they don't give much protection. Besides, mud can grab the sneakers, ripping them from your feet. It's a little trick nature likes to play. On the other hand, hiking boots can be clumsy and therefore inefficient for rock climbing within the cave. A good compromise would be paratrooper boots; they are flexible and have solid soles. Some cavers have been known to ventilate them by cutting holes in the uppers, to permit any water that flows in to flow out again. This alone, however, does not solve the blister problem.

That answer is found through multiple pairs of socks that can take up any slack, preventing the boot from shifting on your foot. Usually, it is sufficient to wear a pair of cotton athletic socks next to your feet, covered by a pair of heavy wool socks. The wool will wick the water away from your feet, in addition to alleviating the slack problem.

If your feet are especially tender, there are two additional avenues for protection. First, you can buy sheets of moleskin with adhesive backing from your local pharmacist and cut out small pads to plaster over the sensitive portions of your feet. Or, you can go the easy route, and purchase a pair of knee-high nylon stockings. By putting these on before the cotton and wool socks, you have what amounts to a second layer of skin that protectively absorbs friction. Either way, your feet should be better off.

So often, an expedition has been spoiled because one person had developed a bad blister—bad enough so he could not walk. It is times like these that try your patience. There is no real excuse for getting blisters. But, on these occasions, the ambulatory half of the team must calmly stop and carry the hobbled caver out of the hole— perhaps harboring thoughts that euthanasia would be a better alternative.

The bywords to remember in cave apparel are "sturdy, warm and a dry second set." Face it, it's quite likely that you will get soaked to the skin and muddy, to boot. A dry set of clothes may turn out to be a real comfort and a lifesaver for the trip back out of the cave.

If you follow all the above advice, you'll be clothed for basic caving—work that does not require crawling, climbing, ducking or gripping. Your body will be warm, but still vulnerable to bumps, knocks and sharp edges, so you'll require special, protective equipment.

The most obvious piece of protective equipment is a hard hat, or helmet. A plastic hard hat is adequate and should protect you from most bumps and falls. It also affords some protection from falling rocks, canteens and other objects and, while no hard hat can stop everything, most provide appreciably better protection than does bare scalp. Although any well-designed hat will do, outfitting shops sell hats specifically designed for cavers and miners.

A helmet also provides a mounting place for any sort of light you may desire to use, keeping your hands free to hold on for dear life and allowing you to see what you are doing at the same time.

Life, like war, gives medals to both heroes and fools. Caving does the same. The fool's award is given out every time a caver goes down without his helmet. Varying in size and shape, the medal is most commonly found in the form of a dented cranium. From early experience, let us tell you that bumps on the head only confuse a phrenologist. Besides, would you like to wince every time your lover runs her or his fingers through your hair?

Wear a helmet at all times.

Two other protective items should be noted, if you are going to come out of a cave without looking like you met up with a wildcat and lost the battle—gloves and pads.

Gloves are the first, and their assets are obvious: They protect the hands from jagged rocks and rough ropes. Usually, a caver uses an unlined pair of gloves made from a supple leather, often deerskin. Sometimes, however, a heavier pair may be necessary to complete rope work of longer duration that might destroy the lighter gloves.

After gloves come pads, and you'll find your coveralls will be minus knees and elbows after one or two trips if you don't use them. Pads can be purchased commercially as "gardener's pads." These are simply rubber knee pads you strap on, but they are bulky and you will discover that you can't walk in them. A better bet is to sew patches of leather into the knees and elbows of your coveralls.

Just as there are certain types of clothing you must have, there are types you must avoid. Shorts of any kind are the equipment of either the extremely brave or the extremely foolish. They leave the legs exposed to all sorts of protrusions and encourage nasty scratches that could, without proper care, become infected. Any loose clothing also creates problems by inviting entanglement in ropes, tight passageways and even other gear. If you're too warm, take off that shirt or jacket. If you just unbutton it and leave it loose, it could lead to disaster. Finally, even though we have stressed warmth as a major need, down clothing is a poor choice for all but the most exceptional circumstances. Down clothes are bulky and fragile and when down is soaked with water, it loses all of its warming characteristics and becomes a heavy, tedious burden.

Immediate Personal Equipment

Aside from the gear you wear, there are certain other items you will need to lug into the cave. Most of these are for your own personal comfort and safety but there are some items you need for your actual exploring activities within the cave.

Among these items is what could be called a mini-survival kit. All the necessities can be packed into a relatively small water-tight container and carried in a pouch or strapped to your lower leg. Since most caving expeditions never run into trouble, it's likely that you'll never even open the kit, but the items in it will be worth their weight in gold if you do run into a problem.

The following survival kit contains items you can purchase at any well stocked store. The utility of the equipment is obvious. This kit is for emergency purposes only. You would turn to it only in case the rest of your gear was lost or destroyed, and you had to wait until somebody came to rescue you.

Survival Kit

- 3-inch plumber's candle
- water proof matches
- halazone (water purification) tablets
- knife/razor blade
- first aid kit (emergency)
  antiseptic
  bandages
  gauze pads
  adhesive tape
- space blanket

For more extensive use, the following type of everyday first aid kit should be carried in your pack.

Standard First Aid Kit

- five 3-inch gauze pads
- bandages
- antiseptic
- rubber tubing/tourniquet
- ammonia inhalers
- snake bite kit
- 50-foot roll of 1-inch gauze
- knife/razor blade
- three triangular cloth bandages
- inflatable splint
- burn ointment

The use of this gear will be explained in greater detail in chapter 6. However, you must be sure to keep it in good shape. Check it out after every trip; that fall you took may have done more damage than you thought. Discovering a shattered bottle of antiseptic as you're cleaning a gash in a cave filled with bat guano can be a very rude awakening.

Among the other equipment you'll be carrying on your immediate person is your source of illumination. Take this bit of advice to heart immediately: *There is no light at all underground—none.* The only light you have is the one you bring in with you. Take care of it, whether it's electric or chemical.

The most popular light source among cavers is the standard *carbide lamp.* Originally developed for miners, it has been adopted by spelunkers as the most dependable and convenient illumination device, because one charge of the chemical calcium carbide will last approximately four hours. And the lamp is convenient because it requires little maintenance and is very lightweight. The carbide lamp works on a gas generation principle. The carbide is placed in the lower chamber of the lamp and a reservoir of water is held above it. When the water trickles down onto the carbide, acetylene gas is formed. The gas escapes through a jet in the reflector and ignites, burning with a clear yellow-blue flame. It will continue to work, barring problems, as long as the gas is generated.

The carbide lamp does have disadvantages, however, and these must be recognized before you accept it as the caver's messiah. First, if you will be in the cave a long time, you will need to carry extra carbide. You must keep it dry, because carbide and water react to manufacture acetylene whether or not the carbide is in a lamp. Get your carbide wet, let a pocket of gas gather and light a match . . .. Well, we'll leave the rest to your imagination. Also, the carbide residue is extremely toxic. Don't dump it on the floor of the cave, or you'll probably ruin the cave for all others (cavers and wildlife) by contaminating the water. Use a little common sense. Take the residue out of the cave with the rest of the trash. Too many nice caves and too many innocent animals have been destroyed by inconsiderate explorers.

The best way to avoid mid-crawl light problems is to know the various parts of the lamp and what can go wrong. The upper chamber of the lamp (the one that holds the water) is the complex portion of the unit. On the top is a toggle that controls the flow of water into the lower chamber—one direction means less water, the other means more. Also in the top portion is a felt pad that controls the flow of gas to the jet and prevents water from fouling the jet. Between the upper and lower chambers is a large rubber gasket designed to prevent gas and/or water from escaping at the junction of the two chambers. The jet found in the center of the reflector and attached to the top chamber refines the flow of gas into a needle-thin stream. It is no larger than a pinhole and is susceptible to dirt and other clogging factors. Without the tip, the flow of gas is too broad to burn properly and the lamp will not work. Finally, on the top of the lower chamber is a series of threads that serves as the connector between the two chambers. They are tricky and you can cross threads if you are not careful.

But what happens if you are deep into an intriguing crawlway and your lamps dies for no apparent reason? A number of problems could be the cause, and the following chart should help troubleshoot some of the major glitches.

Problem	Check	Cure
no flame	water valve adjustment	increase water flow
no flame	lamp bottom	remove and rethread, making sure threads are seated properly
no flame	gasket	replace if worn, clean if needed
no flame	felt	replace if hard or brittle, dry if wet
no flame	tip, for dirt, damage or loss	clean or replace
excessive flame	water adjustment lever	turn it off, wait for excess gas to clear

| weak/angled flame | tip, for partial clog | clean or replace |
| bottom very hot | water adjustment valve | turn it off, wait for excess gas to clear |

There are many more problems and, of course, you could be out of carbide. Check your instruction booklet for further solutions—*before* you begin the journey. A book won't be much help in an unlit cave. On the whole, you should find that carbide lamps deliver excellent light in almost all situations. If, however, you don't seem to be getting enough light, and the unit is working well, two other areas could be causing the trouble. Either your reflector is dirty (clean it!) or, if you are still underground and you can't see the walls or ceiling, you just need more light. You're liable to be spending a lot of time exploring that new room.

Many cavers are now using electric headlamps, but they pose two problems. First, the batteries of the unit have an unmerciful tendency to wear out after four or five hours of use, which leads to the second problem. If you get more potent batteries, you also will end up with a heavier unit. Besides, few units are waterproof and water is the main nemesis of electric current. Get a carbide lamp; you'll be happier.

Your best bets in carbide lamps probably are the Premier or Justrite units. Both can be purchased at most outdoors shops for about $20. A two-pound can of carbide costs about $5, and a cleaning kit for the lamps can be obtained through either the store or the manufacturers. One word of warning: If you will be caving overseas, remember that foreign carbide is much coarser than its American counterpart and it comes in bigger chunks. You may have to grind it yourself to keep your lamp working properly.

Other miscellaneous pieces of equipment you'll need include: a water bottle (for your personal supply), a watch, a compass (for use with a map if the cave has been charted), a knife (useful for quick excavations and cutting slings and clothing free from obstructions) and *marking pieces* (to find your way out of an unmapped cave.) The

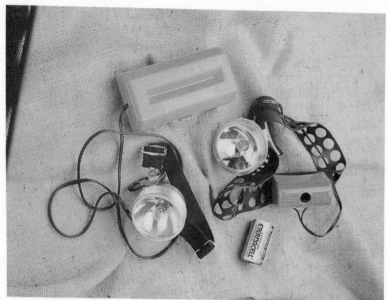

Lithium-powered headlamp. (photo courtesy of Recreational Equipment, Inc.)

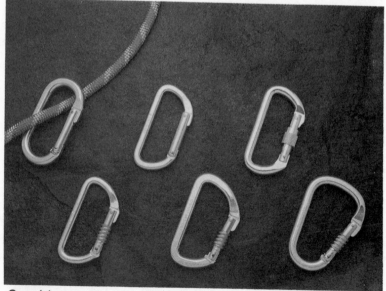

Carabiners: (l to r): Standard oval, standard D, Locking D. (photo courtesy of Recreational Equipment, Inc.)

marking pieces should be either metal arrows pointing toward the entrance or reflective tape, again in arrows. Both must be removable to protect the cave from unnecessary permanent damage.

Climbing Gear

The final bits of gear are your climbing tools: carabiners, *ascenders* and rope.

The simplest description of a carabiner is a link of aluminum with a snap-lock gate. It clips you to your rope, your climbing harness or anything else, for that matter. Used in groups of three, carabiners can act as a *rappel* unit to control your descent and, used alone, one can hold your climbing harness to the rope, allowing you to move up and down in a vertical, rather than a horizontal plane. Carabiners used with brake bars also make an effective rappel control, allowing you to control your speed without undue physical stress.

A carabiner usually is aluminum or an aluminum and chromium alloy. Most test out with a minimum breaking strength of 3,880 pounds, and a maximum of 5,000 pounds. Thus, you should feel pretty secure in using carabiners since you'd probably have to fall some 200-plus feet to create enough stress to break one, and you would be stopped by your belay line well before that.

With carabiners, we've mentioned climbing harnesses and these are simply loops of webbed material of sufficient size to go around your chest or hips. Usually made of one-and-a-half or two-inch nylon, one harness is sized to fit around your chest and snuggly under the armpits. The other actually is a large loop which, when wrapped around your waist and pulled up between your legs, forms an effective seat. The uses of both will be discussed in a later chapter.

Ascenders have a singular use—getting you up once you have gone down—and you need two (sometimes three, depending on your rigging) to climb the rope effectively. Ascenders remove the need for bulky ladders and extremely hazardous hand-over-hand climbing. They work

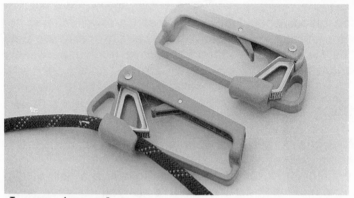

Jumar Ascenders.
(photo courtesy of Recreational
Equipment, Inc.)

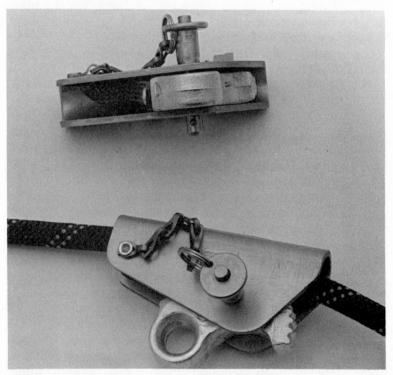

Gibbs Ascenders. Note the geared
cam.
(photo courtesy of Recreational
Equipment, Inc.)

on the eccentric gear principle; that is, as tension is applied to one side of the gear, the other side grips the rope and prevents slipping.

Two types of ascenders are currently in vogue. One, the Jumar Ascender, retails for approximately $65 a pair. However, it uses many moving parts and the more parts, the greater risk of equipment failure. Another type, the Gibbs Ascender, works on the same principle as the Jumar, but it costs only about $55 a pair and has but one moving part. You decide.

When discussing ascenders, you inevitably end up discussing rope. While rope is not an essential personal item, it's always wise for you to bring your own because somebody else is bound to forget his. Usually, at least two ropes, and possibly more (each cave is likely to have more than one drop and ropes must be left in place) are needed by a caving party—one for climbing, one for belay. The length of these lines should be 100 to 120 feet and, if you need a longer rope, you should not be in that cave until you are more experienced. As it is, you're still talking about a 10-story building when you approach a 100-foot drop.

Most cavers use a *dynamic* rope of some sort—"dynamic" meaning the rope is made of a core of artificial fiber surrounded by a sheath of the same fiber, but this time it's woven. While this unique construction gives the rope good strength, cavers should not use a simple dynamic rope, but one that has a high tensile strength, with low stretch qualities. (The rope does you no good if you slip into a 12-foot hole on a 10-foot rope and hit bottom.) A line with these characteristics and especially resistant to mud and water has been developed by a Georgia firm, Bluewater Ltd. The Bluewater II and III lines were developed specifically for caving conditions and will not only serve safely as your main line, but will resist excessive water, mud and abrasion, as well.

However, rope, like all things, wears and eventually becomes unsafe. Watch for frayed or otherwise damaged sections. If a portion of the sheath looks damaged or worn, melt back a portion of it with a small flame and examine the core fibers for damage. If more than 60 percent of the core fibers are damaged the rope is unsafe; *get rid of it* You'll find a new rope costs less than a hospital room these days. Given that rope *does* wear out, however, there are ways you can prolong its life. First, avoid setting the rope through mud or water. No matter how good the line is, in time those two elements will destroy it. If it does get dirty, wash it carefully in lukewarm water and a mild detergent and hang it out to dry. *Never dry the rope in an electric dryer*, if you do, it will shrink and lose strength. Next, if your rope is set so it crosses a rock lip or an outcrop of stone, try to pad the critical area. If you minimize chafing, you'll lengthen the life of your rope.

Other things that inevitably kill a rope take the most unassuming forms—ketchup, vinegar, salad dressings, bleach, household cleaners and carbide residue. Keep your rope away from these substances as much as possible. Also, try to avoid all heat above 180° F., including sunlight. If you leave a rope in the back window of your car, it probably will bake, causing the core fibers to change their characteristics and lose tensile strength. One final rule on rope: Be nice to your rope and it will last long enough to get you out of the cave. If not, it may take revenge.

Group Gear

If you are in a larger caving group—four or more people—there are several pieces of equipment you may find useful in enhancing your comfort underground.

One of the best ideas to come down the pike is the inclusion of a *mini-stove* in your caving supplies. No longer will you have to eat cold rations. Face it, there's nothing better than a hot cup of coffee, tea or chocolate to warm you after a two-hour trek through frigid water. Hot food also tends to lift the spirits and give you some energy reserves. Most stoves use either white gas or kerosene for fuel. They weigh a pound or so and generally stand about five inches high by two inches square—small, light and easy to use. The more advanced stoves (though not necessarily the more expensive) don't require any pumping; you just hold the fuel tank in your hands until the gas volatilizes and begins to flow. Light it, and a'cooking you will

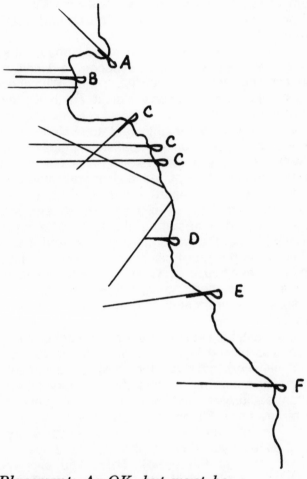

Piton Placement: A: OK, but must be well placed or it could slip. Might be dangerous due to proximity of fractured beds (B). B: NG due to badly fractured section. C: Also bad as the prominence is badly cracked and could disintegrate. D: Also no good as the joint ends and the piton is not driven in far enough. E: An excellent location, but it is not driven in far enough. F: Very safe.

go. One of the better units is the Optimus 8R. It can be bought at any camping supply house for $35 to $40. It can boil water in three or four minutes and heat a can of stew in five.

Of course, if you take a stove, you should take some cooking gear. Pack a fork, a spoon, a small pot (one quart) and, possibly, a small frying pan. With these utensils, you probably can cook almost anything short of a roast round of beef.

Other goodies you might want to include in a group pack include a *Brunton compass* for mapping new passageways, but it is a little exotic in its use and applications and should be acquired only if you intend to use it. Also, if you are on a more scientific bent, a shockproof thermometer would be an excellent investment. Many speleologists record temperature fluctuations in an effort to learn more about the workings of caves. And if your group is proficient in the art of climbing the more technical caves (such as Schoolhouse in West Virginia), special climbing gear may be included. Again, this equipment is not needed unless you are going into a cave marked, "Experienced Cavers Only!"

Pitons (metal spikes to be driven into cracks), *chock blocks* (also placed in cracks, but without driving them in with a hammer), pulleys (for moving equipment across pits) and hammers all fall into the category of special-application equipment, and you really don't need them for the average cave. Pitons are used to provide clip points to attach your rope as you climb a wall. About every 10 feet in your ascent, you drive a piton into a crack and clip a carabiner through the eyelet. Then, you clip your rope through the carabiner. If you fall after you have passed the piton, you will fall only twice the distance of the rope above the piton (i.e., if you are five feet above the piton, you will fall 10 feet).

Chock blocks serve the same purpose as pitons, except that you insert them in the crack and pull them snug. In fact, for aesthetic purposes, they are better than pitons, because you can remove them as you leave the cave. Unfortunately, chocks, like pitons, are dependent on one thing—cracks. Those wonderful splits in the rock are not

common sights in caves since weathering is more the result of water erosion than the expanding and contracting
caused by freezing and thawing. Certain tree roots inspire
cracks, but they cannot be depended on. Instead, check
into chapter 5 for climbing techniques.

Now that you have all this equipment in one pile, you
are faced with the horrible question of how to lug it into
the cave and back out again. The answer is a *Gurnee Can.*
These cans are water-tight containers designed to with-
stand most shocks, and they are bullet-shaped so they
can be pulled through a passageway. Most caving supply
houses can provide your group with Gurnee Cans at a
moderate cost.

Exotica

This section will be short for one reason—most of this
stuff is not necessary to anyone but the most dedicated
caver who makes a habit of going into unusual caves or
on caving expeditions of long duration.

Among the special goodies an expedition-level caver
could use are a two-way radio, an *ice axe, ice pitons, drills*
and *expansion bits* and *inflatable boats*. All of these
pieces of gear have specific applications under special
conditions. Radios are nice if you want to communicate
with a base camp or cavers sitting on the lip of the pit you
just descended (saves the vocal cords). However, CB radio is limited by line of sight; that is, you must have a relatively clear shot at the radio you are trying to contact. In
addition, mineral rich rock walls can act as reflectors,
bouncing the radio waves around the cave, creating radio
interference. Still, it's a nice idea, and worth the try if you
have the extra room to carry a walkie-talkie.

Ice axes and ice pitons tend to belong more in the
world of mountain climbing than in that of caving. However, there are some caves in Washinton, New Mexico and
Idaho where you will find substantial amounts of ice. But
ice is a very fragile form of cave beauty and should not be
marred by either of these two pieces of gear unless it's
absolutely necessary. The ice caves of Mt. Rainier in

Washington may be the only place you can use these pieces legitimately since they provide the only means of reaching them.

Expansion bolts and drills fall into the same category as the ice tools. They damage caves and should be used only when absolutely necessary. The drill is used to create a hole (instead of a crack) so that you can insert a bolt to be used like a piton. It's rough, tedious work. Limestone can be very hard, and it may take an hour or more to set one bolt. One hour for 10 feet—is it worth it? Not really, but it may be necessary to use bolts to open new trails within a cave. Again, though, remember that this type of work should be reserved for expeditions led by competent speleologists.

Inflatable boats also are a nice convenience for the caving group save for two faults: They are heavy and bulky, and they tear easily. Pulling one into a cave only to discover that a rock outcrop got to it first can be very disappointing. If you take one, it is best to load it into a Gurnee Can and hope that you don't kill it while you are using it.

There is much more equipment you can add to your list, but then again, this is meant only to be a guide. As you begin spelunking, you will know better than we what you should take along.

The Cave Gourmet

Now you must determine what food to take along—after all, you have to eat. The first question to be answered is how long you will be in the cave. Add to your answer a margin of time in case something unexpected happens — like a passage that seems to go on for miles. For example, if you expect one meal underground, plan for two. The food should be light weight, easy to prepare and durable. A banana carried in your pocket is likely to resemble a tube of toothpaste when you sit down to eat it. However, you can have your fruit and eat it, too, by simply placing it in your cooking pot. In there, it should maintain its structural integrity.

Originality and creativity are the watchwords in cave

cookery. Canned meats, breads and other foods are available at any supermarket and, with a little work, they can be transformed into quite passable meals. Canned fruit makes a nice break from the ordinary and gives you the energy needed to keep plugging along.

To keep yourself going between meals, many handy snacks can be prepared for use on the trail. However, some types of "squirrel food," also known as "nature food," sold in stores have the texture and taste of dried bat guano. (Don't ask us how we know.) It's always better when you make it yourself, and things like *gorp* (a mixture of nuts, pretzels and cereal) can be pretty tasty. These, however, require water to wash them down, so it's better to use dried fruit instead of pretzels in your gorp recipes. While you will still need water, you might not need as much. Whatever you eat, however, you will still need salt. Since you will sweat, you will lose much salt and, if you do not replace it, you will become ill. Salt tablets will solve this problem. For quick energy, and a welcome change, you can buy honey in a tube. Since it is predigested by the bees, honey requires no lag time before it is digested by your system, so you will absorb the sugar almost immediately after you swallow the honey.

Chapter 3

Getting Ready

Now you have all that beautiful equipment and are ready to go caving. Well, the sentiments are great, but if you go caving at this moment, you will never want to go again—either because you are miserable or because you got hurt and had to be rescued. You should no more go caving without proper physical conditioning than you should attempt to compete in the Boston Marathon without training. In fact, many caves are not unlike a marathon course, placing seemingly impossible demands on your stamina and strength. In some caves, you never get a break, so you have to be prepared to go the distance up to 12 hours or more.

Caving does not require the use of any specific set of muscles as do swimming or running. Rather, cavers must be in the same shape overall as athletes participating in football or basketball. Every part of your body will be called upon to serve—arms, legs, fingers, back, abdomen, hips—even toes. Indeed, cavers take the admonition, "Be on your toes," quite seriously.

You would be amazed at what you are called upon to do in the slimy confines of a subterranean passage, and the importance of all-around conditioning is not to be underestimated. It could make the difference between a pleasant experience and tragic consequences. When you are underground, you literally and figuratively must be able to pull your own weight because in many situations you must be responsible for getting yourself out of the mess you may have gotten yourself into.

Throughout this chapter, we will discuss various methods of training and the situations they are designed to meet. At times, the descriptions may seem imprecise. This is because each person responds differently to similar situations. You may even know of other techniques which will better suit you. All you need to remember is that you must be in good physical condition before you go in.

There may not be anyone around to go in and help you out.

Aside from teaching you how to walk with maximum efficiency, hiking is a good way to develop and test your conditioning. Schedule several hikes before you attempt your first cave. Part of the hike should include a jaunt through swampy conditions, especially if you intend to explore any wet caves, and crawling through some good-size culverts also should be included. It's a good idea to keep hiking, even after you have begun caving. Those long, outdoor walks do much to keep you in condition, and the more you hike, the higher you can develop your endurance levels. Do not, however, try to substitute hiking for learning good caving technique. Endurance will do you no good if you can't move around with relative ease.

Reptation

A wise old caver once said "A man must rept before he can climb!" You are probably more familiar with the old "learn to crawl before you walk" adage, and *reptation* is just a fancy word for crawling. The beautiful rooms found in American caves are connected by almost impossible crawlways, and if you are to be a successful caver, you must know how to crawl without wearing yourself out. At times, crawling can vary from the traditional and relatively easy hands-and-knees method to slithering along like a snake, moving only inches at a time. To illustrate for yourself the difficulties of crawling, take a few dining room chairs and set them up around a room. Next, take some water glasses and lay out a winding course between the chairs. Now, comes the fun part, and you could make it even more interesting if you fill the glasses with water. Crawl between the chair legs and through the course of glasses, without knocking over the glasses or moving the chairs. Now imagine that these chairs and glasses are only the visible part of 200 feet of solid rock. Easy? O.K., then crawl under your bed without moving it!

The object of the crawling exercise is to show you just how difficult this seemingly simple movement can be. To squeeze through some really tight spots, you probably will find it necessary to thrust your arms past your head. This will lower your profile and enable you to get through slots as narrow as fourteen to eighteen inches, and these little beauties are made just for the slithering action of true reptation. Once beyond the basic crawl, reptation is just what it sounds like—a snake-like motion allowing the caver to gradually inch along apparently prohibitive slots and passages. Many European spelunkers prefer to do this in the buff, but rock can be awfully hard on the skin. However, it may be necessary for you to do just that—strip down—to get to that fantastic room just 20 feet away. Remember, the true caver has no shame.

Exercises to improve crawling muscles include the traditional situp, pushup and squat thrust, to increase the strength of your abdominal muscles, upper arms and thighs. You will however, find that these exercises alone will not fully aid you in your efforts to crawl with ease. Many times you will have to depend on your toes to give you that extra bit of power to slide through an especially tough spot. Thus, ballet toe exercises are very valuable. Practice lifting yourself up and down on your toes for about five minutes at a time. After one round of this it is guaranteed that you will want to drop, but you will, with continued practice, develop the strength and flex in those seemingly useless digits to be able to force your way along a passage. In some cases, your toes will even provide the prime motive force.

Another major aid in crawling is breath control. Although, at times, it will be next to impossible to breathe properly, you can conserve your precious oxygen, and not hyperventilate in the process, if you learn breath control. Basically, this is a way to relax your torso and draw air into the lungs using only your diaphragm. Also, when your gut is hard up against a cave wall, you will be able to breathe without feeling like you are being crushed.

If you practice hatha yoga (a good conditioning regimen in itself), you already know how to control your breath. Yoga, however, is not a requisite for caving, but is an alternative and easy method of developing breath control while you take a short break during your exercise period. While you are still huffing and puffing, lie on your back with your arms over your head. Stretch. Then take a slow, deep breatn. If you gasp it in, exhale and try again. Eventually, you will be able to allow your body to relax, even though you are in the midst of strenuous physical activity. By stretching your arms over your head, you elongate your chest cavity, thus making it easier to take in more air without expanding your chest as much as in normal breathing. And with additional air you get more oxygen, and you will be negating some of the danger from exhaustion.

Another exercise that will help you crawl is known as *seal-walking*. This is an exercise very popular among marine-recruit instructors in boot camp. Cradling their rifles in their arms, the marine trainees "walk" on their elbows using this as their only method of locomotion across a field. You can do it in your own home, sans rifle and drill sergeants, to great benefits. Just be sure not to use your legs and, after you have treated your rug burns, you will begin to appreciate the way you have developed your upper arms and chest. Even more so after you enter a cave.

Walking

While caving usually is a series of variations on the theme of crawling, there are times when you will be able to walk along broad avenues. Learn to rest while you saunter along. Pace your steps so that they require the least amount of extra muscle action. Lift your knee and walk with your whole leg, not just from the knee down as many people have fallen into the habit of doing. If you do not walk correctly, you could be wasting valuable energy, making the return trip that much harder.

Incidentally, jogging will not teach you how to walk or how to conserve energy. It is, however, a valuable aid in improving your stamina and increasing blood flow.

Climbing and Descending

Many people make a serious mistake when they assume the arms contain the major climbing muscles, content to drag themselves up pitches and cliffs. Quite the contrary the legs are the main thrusters for any climbing efforts that are to succeed without super-strain, pushing, rather than pulling yourself up. Remember that your legs are capable of lifting hundreds of pounds—they are exercised almost every waking minute. Your arms, on the other hand, generally are not capable of lifting more than one hundred pounds. The difference is apparent. While you may be able to haul yourself up a pitch of 20 feet with your arms, don't count on doing so on the next one.

An excellent way to condition your legs and arms for the rigors of climbing is to initiate a strenuous weight training program. Once or twice a week, spend an hour lifting weights in various exercises. The bench press will strengthen your arms, while squatting with weights on your shoulders will build your thigh and calf muscles. There are, of course, other weight exercises for other parts of your body, and if you buy a set of weights, the instruction booklet will have detailed instructions and explanations of the direct benefits of each. The same information and the weights are available at your local gym or health club.

Eventually, you will be able to increase the amount of weight you use on each lift, but an old rule of thumb for balanced muscle development is to repeat lifting a moderate amount of weight, rather than trying to hoist your maximum. This not only prevents you from becoming a Mr. Universe candidate, but also is a far safer method of exercise. However, if you really want to increase your muscular strength, do not stop the weight exercises when you feel tired; go that one last lift to rupture your muscle tissue. The key to successful muscle development through weight training is that Mother Nature tends to overcompensate when repairing damaged tissue. When your muscles heal, the result will be more and stronger tissue than when you began. And, if you use the weights only twice a week, the added muscle will not be apparent until you need that extra strength. A regular program of this sort, then, will enable you to demand more from your body— and get it.

Generally, you should strive to build your entire body. Sports such as handball, tennis, squash and even swimming will allow you to develope your muscles and stamina while having fun and keeping limber at the same time. Never forget that big muscles get in the way in little passages.

Physical Limitations

Some of us are blessed with small bone structures and some are not. In caving, the small guy invariably is the one volunteered to test a small passage. If he can make it, the next larger tries and on and on until somebody gets stuck. One such case that comes to mind occured while caving in New York state. In Knox Cave, south of Albany, is a passage called the Gun Barrel. The reasons are obvious. The passage is 75 feet long and only inches in diameter. Most of the group was able to go through, but there were five of us left on the other side due to broad shoulders or hips. We never did get to see the beauties of the Alabaster Room with its pure white formations.

On the bright side, neither did we end up trapped on the inside when the lead man became wedged upon returning. The five of us had to go for help when it proved impossible to extricate him, and for 10 hours police and firemen labored to get him loose. Finally, when all else seemed lost and we were getting ready to call the special rock drilling team from Renssalaer Polytech, he was pulled free. Even large size has its advantage in caving.

Aside from these occasional problems, no physical limitation should prevent anyone from caving. However, caving does require complete use of all hands and arms available. Do not attempt a cave if you are tied down with a broken arm or worse. They are just so much dead weight. In addition, good vision is a must and should not be impaired by anything—liquor, sweat bands, whatever.

There is but one problem that might prevent you from fully enjoying the world beneath the surface, and that is *claustrophobia.* Don't laugh. When you are being pushed into an ever tighter slot, even the most logical mind can play nasty tricks. You will feel like you are being crushed, like the walls are closing in on you. As the going gets tougher, not only can the tough get going, they can get panicked. Under the latter circumstance, you may do something very foolish and end up in a worse state than you began. While there is no way to avoid that feeling, you can minimize the psychological effects by reminding yourself that there are other people nearby who can help you if a problem arises. But, above all, *keep your head*!

Fear is something nobody can be conditioned to avoid. However, the onset of fear can be prevented if you observe a few simple rules. First, get in topnotch shape before you go caving. That way, you will be able physically to surmount any difficult section that may appear. Next, be confident of your abilities. You *can* do what you attempt, provided you do not attempt anything beyond your limitations. Finally, never succumb to the wiles of fright. Panic will lead you into more dangerous terrain, and if you keep your cool, there is little that can harm you.

Both mental and physical conditioning, then, are important factors in any successful caving expedition. Remember, a cave party is only as strong as its weakest member. If you are out of shape, you will spoil the day for everybody, including yourself. Also, you are courting disaster by not being prepared for the rigors of the journey. If you are not able to enjoy the trip, if you are always struggling to keep up with everybody, you probably will never go caving again. This is not to say that novices are unwelcome; we all had to start somewhere. Bad memories, however, are a strong reason to stay away from caves, and there is no need for you to create bad memories by getting worn out and hurting yourself. Remember, there is joy through strength.

Chapter 4

Now That You're Ready . . . Where Do You Go?

You've bought your equipment, test-hopped it, fallen down a few times in practice, and you generally feel ready and rarin' to go. Now just where on (or under) God's green earth will you find a cave worthy of your efforts? The search should not be too arduous since (as mentioned earlier) almost every state in the U.S. has at least a few caverns waiting for a caver to open their mysteries. Prime caving country often is only a few hours' drive from a major metropolitan area. Many, however, if not all, caves are located on private land or, in the case of the Great American West, on restricted government holdings. In this chapter, we offer some essential caving precepts from the realm of diplomacy, to help you find your own caving experience.

Old Mac Donald Had a Cave

As you cruise down Highway 92 following the tortuous directions passed on to you by a fellow spelunker, you discover that the cavern you seek is on land posted every two feet with forbidding signs exhorting trespassers to stay clear. What to do?

More often than not, the farmer in question has been bothered by cavers, hikers and other motley adventurers to such an extent that he does not want a single soul entering that cave. Usually, he has allowed cavers on his property in the past and they have abused the privilege by defacing his land. The old parable of "One Guy Screwing up a Great Situation for Everybody" holds true in caving.

Remember, the owner of any land you'd like to explore is under no obligation to you. He is not required to let you into his cave. If he has not been too sorely abused in the past, however, an owner usually is willing to allow small groups of responsible cavers onto his land for the sole purpose of going into the grotto. All you have to do is ask for permission. A special note: In the case of more cagey farmers (especially dyed-in-the-wool Yankees), a little financial incentive sometimes is necessary.

Besides considering the owner's feelings and property rights, you also must remember that the entire community around the cave is affected by invading explorers. Making a mess of local public facilities, leaving more mud in a laundromat than the whole cave contains or raising hell in a restaurant may damage relations for future spelunkers as much as trashing the cave itself. The ire of an outraged community can inspire the placement of concrete blocks in a cave entrance pretty quickly. Be as considerate while visiting a town as you would be in your own home town.

Beyond simple courtesy lies the world of interpersonal relations. Get to know the owner when you go to see him about entering his cave. Learn about his family. *Listen to him.* That fantastic tale about his grandfather's explorations may contain an important link to the cave itself. Old Mac Donald also may tell you about some unforeseen hazard in the cave that you will appreciate later. Once you take the time to listen to such individuals, you will begin to appreciate the difference between your life in the big city and their life in the country. You may even learn something about people. Caving is more than just climbing and slogging through the mud.

Once you have obtained permission to enter a cave, you have to get to its entrance. Since you have probably driven to the site, it is important to ask the owner where you can drive and park your car. As a city dweller, you may not immediately recognize the damage your car can do to crops and livestock. Too many owners have been antagonized by cavers blithely driving their autos through a field of grass that turned out to be alfalfa. Crop damage hits the farmer where it hurts most—in the pocketbook. Also, there is no greater irritant to a farmer than tire ruts across land that originally had no roads. Be smart: *Stick to the route the owner gives you.*

On the whole, property owners welcome cavers, as long

as they can prove that they are conservation-minded and they recognize that they are guests on the land. Only one thing will instigate bad feelings—evidence of foolishness. The owners know (as we do) that clowns probably will get hurt in a cave, and if there is an accident, the farmer's land will be devastated by the subsequent rescue effort. Of course, the property owner has every right in the world to tell you to stay off his land, to get off his land or to leave his cave even if he has previously given you permission to explore. He can have you arrested for trespassing as well, and working on a chain gang in Mississippi may not sound like the perfect end to your adventure. Be considerate, be smart.

Good will is a tough commodity to generate in today's world. If you find a friendly farmer, be nice to him. Do not abuse the privilege he has granted to you by foolish actions of a destructive nature. Not only will you lose a friend, also, you will damage the reputation and credibility of every other caver the world over. Heed the lesson learned by contrite snowmobilers who forfeited the privileges granted to them by leaving smashed foliage and trash in their wake. Treat the farmer as a friend and he will extend the same respect to you.

Where to Go?

Prime caving country can be found with many forms and features and, within close range of major metropolitan areas, caves can be found in all sizes and shapes.

The following is a rudimentary list of caves and caving terrain within close proximity of major U.S. cities. For more detailed information, contact local Chambers of Commerce. Usually, they are happy to send any data that will bring more revenue into the town. Other excellent sources of information are the grottos, or local chapters of the NSS. To contact the grotto near you, see chapter 8.

New York City/Boston/Philadelphia/ Washinton, D.C.:

Knox, N.Y.—About 20 miles west of Albany on NY 146.

Cobleskill, N.Y.—Secret Caverns, Howe Caverns and others; I 88 to Cobleskill.

Great Barrington, Mass.—U.S. Rt. 7, 13 miles south of I 90.

Bethlehem/Allentown, PA. — Lost River Caverns and others.

Reading, Pa.—Onyx Cave, Crystal Cave and others; U.S. Rt. 222 north to Moslem Springs, Pa., and Pa. Rt. 143 west.

The entire Shenandoah River Valley in Virginia and West Virginia—I 66 west from Washington, D.C., to I 81. Travel south or north. An alternate route would be to take I 66 to Front Royal and U.S. Rt. 340 south. This route will take you to Luray Caverns and many others. (Many cavers insist that the caves of the Shenandoah are among the most beautiful in the world.)

Pittsburg/Cleveland and Virginia:

West Virginia—An area bounded by Interstates 77, 64 and 81 contains some of the nicest caves around. Some of the town names indicate possible caverns or cave-like formations (i.e., Sinks Grove, Salt Sulphur Springs, Sinking Creek).

Southern Ohio—U.S. Rt. 33 southeast of Columbus takes you to the town of Gibisonville (exit south on Ohio 180). The area is permeated with small caverns, including Old Man's Cave and Ash Cave.

Northern Ohio — I 90 west to Toledo and I 75 south to Findlay. In Findlay, take Ohio 568 east about 10 miles to the area of Indian Trail Caverns. Many small caves can be found along with the larger parent cavern.

(Note: A band of cave formations runs east to west across most of Ohio on a line from Steubenville through Columbus to Sidney.)

Raleigh/Winston-Salem/Charlotte:

The Blue Ridge Mountains (North Carolina and Tennessee) — Next to the caverns of central Kentucky, probably the most prolific and extensive in the United States. Take I 40 west to Asheville, N.C.,

Nature did not work a four-day week or observe holidays during the formation of Luray Caverns, subterranean natural in the scenic and historic Shenandoah Valley of Virginia.

According to scientists, small drops of water working day and night over a period probably exceeding 10,000,000 years, created this natural wonder, now enjoyed by more than one-half million visitors annually. Luray Caverns Onyx builds at the very slow rate of one cubic inch in 120 years.

and then U.S. Rt. 23 north toward Johnson City,
    Tenn.
Detroit/Chicago:
    The best bet for both of these metropolitan areas is to
head south into southern Indiana or central Kentucky.
Both areas abound with the most incredible caverns in the
United States.

From Indianapolis — Take I 65 south to Kessinger,
    Ky. (about 90 miles south of Louisville). Here you
    will find the Mammouth/Flint Ridge system, Hidden
    River Cave and Diamond Caverns, to name just a
    few. Southern Indiana — Take I 65 to New Albany
    and 64 west to Tower, Ind. In Tower, take Indiana
    Rt. 66 south a few miles to Indiana Rt. 62/66 east.
    Take 62/66 east to Wyandotte Cave.
Also of interest to the caver is the fact that the southern
counties in Indiana — Orange, Crawford, Harrison and
Perry — literally are peppered with sink holes and other
cavern indications. Good prospecting territory.

Southern Illinois — The Carbondale area is mostly
    limestone and is good caving territory.
Northern Illinois — The stretch of the Mississippi Riv-
    er between Moline and Galena offers several good
    caving opportunities.

The Mississippi River valley also provides us with some
interesting caving experiences. Few caverns in the area
are extensive, but most feature nice formations. Spelunkers in the states along the river are never very far from a
cavern. Also, the great limestone formations of central
Kentucky and Tennessee swing across the river between
Memphis and St. Louis, giving Arkansas and Missouri a
great number of caves.

As one moves west, cities and caves are spread farther
apart. The best caving terrain can be found in two bands
—one in the Badlands of North Dakota, South Dakota,
Montana and the other running across northern New Mexico and Arizona, through southern Nevada and into California.

Los Angeles/San Diego/Las Vegas/San Francisco:

Grand Canyon Area — Take I 40 to Kingman, Az. Take U.S. Rt. 66 east to Truxton (about 40 miles). In this area you will find Grand Canyon Caverns as well as pueblos and other cave formations.

Las Vegas Area — Off I 15 east of Las Vegas are a number of caves, many with entrances visible from the highway. However, a word of warning is in order: Much of the area is out of bounds due to nuclear testing. Find out which sites are O.K. before you start trudging. Also, remember that distance can be deceptive. Those mountains are huge and what seems to be three miles away may be ten.

San Diego — Littoral Caves abound on the coast north of the city.

Sacramento/San Francisco Area — Sierra National Forest and Yosemite National Park abound with spectacular caverns. Contact the U.S. Forest Service for information on open caves. Also, just west of the parks in Calaveras County are some of the more intriguing caves in the U.S. (Mercer Caverns, Moaning Caves). The area is ripe for caves and more are being discovered every day. Also, north of this area is the Lava Beds National Monument, rife with lava tube formations.

Portland/Seattle:

Bend, Oregon —Take I 5 south to Albany. Take U.S. Rt. 20 east about 130 miles to Bend. Off U.S. Rt. 97 south, you will find the Arnold Ice Cave, the Lava River Cave and other lava tube formations.

Mt. Hood Area — Chock full of lava tubes.

(*Note:* Most of the caving in this area is to be found in the lava tube systems just being opened by vulcanospeleologists.)

The above list is just a starter for the enterprising tyro. A good way to find more information on caves in your area is to drop a line to the National Speleological Society, Cave Avenue, Huntsville, Alabama 35810.

For those who are not quite ready for the big jump, there are some 200 caves that are open to the general public and are located on public land (for example, Carlsbad Caverns, Wind Cave in South Dakota, or Mammoth Cave in Kentucky).

While you may not be able to do any actual spelunking in one of these "tourist" caves, you can enjoy the casual type of caving that does not require special equipment (other than a light jacket) or exertion, beyond climbing a steep flight of stairs. Most commercial caves are limestone, but they offer an opportunity to observe the full range of limestone formations in absolute safety under the direction of an experienced guide.

With this list, you can plan excursions into some of the commercial caverns that span the country.

Included herein will also be some undeveloped caves, designated with a "W" for "wild."

Identifying marks:
SR: State Route or Highway   V: Lava Tubes
US: United States Route   G: Gypsum Caves
L: Littoral Caves   S: Sandstone Caves

Frozen waterfall, Green Lake Room,
Carlsbad Caverns.
(photo courtesy of Carlsbad Caverns
National Park)

ALABAMA
BASS CAVE: SR 117 near Russell Cave
CATEDRAL CAVERNS: US 72 in Grant
CRYSTAL CAVERNS: County 30 14 miles northeast of
    Birmingham
GUNTERSVILLE CAVERNS: SR 79, 8 miles north of
    Brooksville
MANITOU CAVE: off US 11 in Fort Payne
NATURAL BRIDGE: US 278, 60 miles west of Cullman
    (Note: A Natural Bridge is a spelaen formation of in-
    terest to scientists as well as the casual explorer.)
RICKWOOD CAVERNS: SR 31 north of Warrior
RUSSELL CAVE NATIONAL MONUMENT: US 72, south-
    west of Chattanooga
SEQUOYAH CAVERNS: off I 59, 5 miles south of Sulp-
    hur Springs
SANTA CAVE: US 72 west of Scottsboro

ARIZONA
(S) CANYON DE CHELLY: I 40 west of Gallup, N.M.
    COLOSSAL CAVE: I 10, 20 miles east of Tuscon
    GRAND CANYON CAVERNS: US 66, 25 miles west
    of Seligman
(S) MONTEZUMA CASTLE NATL. MONUMENT: SRs 79,
    279, 179 near Camp Verde
(S) NAVAJO NATIONAL MONUMENT: SR 64 near To-
    nalea
(V) SUNSET CRATER NATL. MONUMENT: US 89 north-
    east of Flagstaff
(S) TONTO NATL. MONUMENT: SR 88, 90 miles east
    of Phoenix

ARKANSAS
BIG HURRICAINE CAVE: US 65 15 miles east of Ever-
    ton
BLANCHARD SPRINGS CAVERNS: off SR 14 in Moun-
    tain View
(W) BAT CAVE: Buffalo River State Park on SR 14
    south of Yellville
BULL SHOALS CAVERNS: SR 178 in Bull Shoals
(S) CAVE CITY CAVE: US 167 in Cave City
CIVIL WAR CAVE: US 71 west of Bentonville

DEVIL'S DEN STATE PARK: SR 170 north West Fork
DIAMOND CAVE: SR 7, 5 miles west of Jasper
MYSTERY CAVE: SR 21 5 miles north of Berryville
MYSTIC CAVERN: SR 7 south of Harrison
OLD SPANISH TREASURE CAVE: SR 59 north of Gra-
    vette
ONYX CAVE: US 62 east of Eureka
OZARK MYSTERY CAVE: SR 27 southeast of Harriet
ROWLAND CAVE: SR 14 in Mountainview
SHAWNEE CAVE: SR 14 south of Yellville
WONDERLAND VISTA CAVE: off US 71 in Bentonville in
    Bella Vista Park

CALIFORNIA
BOYDEN CAVE: SR 198 near Three Rivers
CRYSTAL CAVE: SR 198 50 miles north of Visolia
(L) LA JOLLA CAVE: actually a number of caverns
    along the coast in La Jolla
LAKE SHASTA CAVERNS: US 99 20 miles north of
Redding
(V) LAVA BEDS NATL. MONUMENT: off SR 139 20
    miles south of Tulelake
MERCER CAVERNS: SR 4 north of Murphys
MITCHELL CAVERNS: northwest of Essex off I 40
MOANING CAVE: near Vallecito
SUBWAY CAVE: off SR 44 northeast of Manzanita Lake

COLORADO
CAVE OF THE WINDS: on US 24 near Manitou Springs
MESA VERDE NATIONAL PARK: US 160 east of Cortez

DELAWARE
(W) An unnamed cavern;   located off Beaver Valley Rd.,
    on the Pennsylvania Delaware line

FLORIDA
FLORIDA CAVERNS: SR 167 south of Marianna
OCALA CAVERNS: US 441 south of Ocala

GEORGIA:
CAVE SPRINGS CAVE: north US 41 and SR 100 in
    Cave Springs

IDAHO
(V) CRATERS OF THE MOON NATIONAL MONU-
MENT: US 20, 26, 93A near Arco
CRYSTAL ICE CAVES: SR 39 near American Falls
MINNETONKA CAVE: east of US 89 in St. Charles

ILLINOIS
CAVE IN ROCK: SR1 40 miles southwest of Evansville
BAT CAVE: in Mississippi Palisades State Park, SR 84
south of Savanna
BOB UPTON CAVE: in Mississippi Palisades State Park,
SR 84 south of Savanna
(S) FERNE CLYFFE STATE PARK: SR 37 south of Mar-
ion
GIANT CITY STATE PARK: off US 51 south of Carbon-
dale
MATHIESSON STATE PARK: SR 71 southeast of La
Salle
STARVED ROCK STATE PARK: SR 71 near La Salle

INDIANA
CAVE RIVER VALLEY PARK: off SR 60 in Campbells-
burg
(W) BEAR DEN CAVE
(W) CRYSTAL SPRING CAVE
(W) ENDLESS CAVE
(W) FROZEN WATERFALL CAVE
(W) LAKE CAVE
(W) LAMPLIGHTER'S CAVE
(W) RIVER CAVE
DONALDSON CAVE: Spring Mill State Park, SR 60 east
of Mitchell
HAMER'S CAVE: Spring Mill State Park, SR 60 east of
Mitchell
MARENGO CAVERN: off SR 64 in Marnego
PORTER'S CAVE: SR 67 in Paragon
TWIN CAVES: Spring Mill State Park, SR 60 east of
Mitchell
SUNKEN CAVE: McCormick's Creek State Park, SR 46
east of Spencer
WOLF CAVE: McCormick's Creek State Park, SR 46
east of Spencer

WYANDOTTE CAVERNS: US 460 in Wyandotte (One of
the most spectacular cavern systems in the U.S.)

IOWA

CRYSTAL LAKE CAVE: US 52 and US 67 south of Du-
buque
MAQUOKETA CAVE: in Maquoketa State Park, SR 130,
Maquoketa
SPOOK CAVE: US 52 west of McGregor
WONDER CAVE: SR 9 northeast of Decorah

KENTUCKY

(W) BAT CAVE: SR 182 about 10 miles from Olive Hill
CARTER CAVES: SR 182 about 10 miles from Olive Hill
(W) CASCADE CAVE: SR 209 about 6 miles from
SR182 near Olive Hill
DANIEL BOONE'S CAVE: US 27 20 miles south of Lex-
ington
CRYSTAL ONYX CAVE: US 31 W, 7 miles northeast of
Park City
DIAMOND CAVERNS: SR 255 in Park City
(W) HORN HOLLOW CAVE: SR 182 about 10 miles from
Olive Hill
(W) LAUREL CAVE: SR 182 about 10 miles from Olive
Hill
LOST RIVER CAVE: US 31W south of Bowling Green
MAMMOTH CAVE: Access by either I65 from Nashville
or Louisville or SR 255 from Park City (Possibly the
largest system in the world, Mammoth Cave Park of-
fers a variety of tours from short jaunts to an all day
epic.)
MAMMOTH ONYX CAVE: SR 335 west of I65
(W) SALTPETER CAVE: SR 182, about 10 miles from Ol-
ive Hill
(W) X CAVE: SR 182, about 10 miles from Olive Hill

MAINE

(L) ANEMONE CAVE: in Acadia National Park, south of
Bar Harbor

MARYLAND
CRYSTAL GROTTOS: in Boonsboro on SR 34, south of
    US Alt. 40

MICHIGAN
BEAR CAVE: 4 miles north of Buchanan

MINNESOTA
MINNESOTA CAVERNS: off US 16 10 miles southeast
    of Spring Valley
MYSTERY CAVE: off US 16 10 miles southeast of Spring
    Valley
NIAGRA CAVE: US 52 5 miles southwest of Harmony

MISSOURI
BIG SPRINGS ONYX CAVERNS: US 60 west of Van Bu-
    ren
BOONE CAVE: off US 70 near Rocheport
BRIDAL CAVE: SR 5 north of Camdenton
BLUFF DWELLERS CAVE: SR 59 south of Noel
(W) COVE HILL CAVE: US 50 10 miles west of Drake
CRYSTAL CAVE: US 65 north of Springfield
CAMERON CAVE: US 61 south of Hannibal
CIVIL WAR CAVE: east of US 65 north of Ozark
CRYSTAL CAVERNS: SR (BUS) 37 north of Cassville
FAIRY CAVE: SR 13 south of Reed's Spring
FANTASTIC CAVERNS: SR 13 northwest of Springfield
FISHER'S CAVE: SR 155 in Sullivan
HONEY BRANCH CAVE: off SR 14 west of Ava
JACOB'S CAVE: off SR 5, 5 miles south of Versailles
(W) MARCELLUS CAVE: I–44 to SR 8. South 7 miles.
MARVEL CAVE: SR 76 west of Branson
MARK TWAIN CAVE: SR 79 south of Hannibal
MERAMEC CAVERNS: US 66 west of Stanton
MISSOURI CAVERNS: (CATHEDRAL CAVE) I–44 10
    miles southeast of Leasburg
MYSTIC RIVER CAVE: US 54 and Rt. K in Camdenton
OLD SPANISH CAVE: US 65 north of Reeds Spring Junc-
    tion
ONADAGA CAVE: I–44 near Leasburg
OZARK CAVERNS: off US 54 near Osage Beach and
    Camdenton

OZARK WONDER CAVE: off SR 59 north of Noel
REBEL CAVE: near the Junction of US 67 and SR 34
ROUND SPRING CAVERN: off SR 19 in Round Spring
(W) SALTPETER CAVE: 7 miles northwest of Rolla
(W) SPENCER CAVE: I-44 15 miles northwest of Rolla
(W) SMITTLE CAVE: I-44 and SR 5 (22 miles south)
STARK CAVERNS: off US 54 near Eldon
TRUITT'S CAVE: at the junction of US 71 and SR 59 in
    Lanagan

MONTANA
LEWIS AND CLARK CAVERNS: US 10 20 miles west of
    Three Forks
Unnamed ice caves in Custer National Forest near Shriver

NEBRASKA
(S) ROBBER'S CAVE: US 77 south of Lincoln

NEVADA
(W) GYPSUM CAVE: 20 miles northeast of Las Vegas
    off I 15
LEHMAN CAVES NATL. MONUMENT: SR 74 off US 6

NEW MEXICO
BANDELIER NATL. MONUMENT: SR 4 near Los Alamos
CARLSBAD CAVERNS: US 62 and US 180 near Carlsbad
(V) THE DESERT ICE BOX: SR 53 southwest of Grants

NEW YORK
HOWE CAVERNS: SR 7 north of Central Bridge (one of
    the oldest commercial caves in the US)
(W) KNOX CAVE: 25 miles southwest of Albany in Knox
NATURAL STONE BRIDGE AND CAVES: US 9 in Potters-
    ville
SECRET CAVERNS: 35 miles west of Albany near Co-
    bleskill

NORTH CAROLINA
LINVILLE CAVERNS: US 221 north of Marion

*Temple of the Sun, Big Room, Carls-
bad Caverns.
(photo courtesy of Carlsbad Caverns
National Park)*

*Formations in the Dome Room, Carls-
bad Caverns.
(photo courtesy of Carlsbad Caverns
National Park)*

*The King's Palace, Carlsbad Caverns.
(photo courtesy of Carlsbad Caverns
National Park)*

OHIO

ASH CAVE: SR 374 near Logan in Hocking Hills State Park

CRYSTAL CAVE: on South Bass Island in Lake Erie, take the Ferry from Port Clinton

DEVIL'S DEN: in Devil's Den Park, take County 10 from Gnadenhutten

OHIO CAVERNS: SR 245 near West Liberty

OLENTANGY CAVERNS: off SR 315, 10 miles north of Columbus

PERRY's CAVES: on South Bass Island in Lake Erie, take the Ferry from Port Clinton

SENECA CAVERNS: near SR 18 and 269 south of Bellevue

SEVEN CAVES: near US 50 west of Bainbridge

WETZEL CAVE: in Devil's Den Park, take County 10 from Gnadenhutten

ZANE CAVERNS: SR 540 east of Bellefontaine

The Palace of the Gods at Ohio Caverns.
(photo courtesy of Ohio Caverns)

OKLAHOMA
(G) ALABASTER CAVERNS: SR 50 south of Freedom

OREGON
(V) LAVA RIVER CAVES: near US 97, south of Bend
(V) LAVACICLE CAVE: near US 20 near Bend in Des-
    chutes National Forest
OREGON CAVES NATL. MONUMENT: SR 46 southeast
    of Cave Junction
(L) SEA LIONS CAVE: US 101 north of Florence

PENNSYLVANIA
CRYSTAL CAVE: off US 222 near Kutztown
INDIAN CAVERNS: SR 45 in Spruce Creek
INDIAN ECHO CAVERNS: on US 322-422 in Hummels-
town
LAUREL CAVERNS: US 40 east of Uniontown
LINCOLN CAVERNS: US 22 west of Huntington
LOST RIVER CAVERNS: SR 412, 3 miles from Bethle-
    hem
ONYX CAVE: SR 662 south of US 22 in Hamburg
PENN'S CAVE: SR 192 east of Centre Hall
WONDERLAND CAVERNS: off SR 37 west of Bedford
WOODWARD CAVE: off SR 45, 25 miles west of Lewis-
    burg

TENNESSEE
ALUM CAVE: US 441 in Great Smoky National Park
BRISTOL CAVERNS: US 421 southeast of Bristol
CRYSTAL CAVE: US 41-64, 5 miles west of
    Chattanooga
CUDJO'S CAVE: US 25 east of the Cumberland Gap
CUMBERLAND CAVERNS: SR 8 10 miles east of
    McMinnville
DUNBAR CAVE: SR 79 east of Clarksville
JEWELL CAVE: 15 miles northwest of Dickson
    and US 70
THE LOST SEA: US 11 5 miles from Sweetwater
LOOKOUT MOUNTAIN CAVERNS: SR 148 near
    Chattanooga
RUSKIN CAVE: 5 miles northwest of Dickson and US 70

TUCKALEECHEE CAVERNS: SR 73 near Great Smoky
    National Park
WONDER CAVE: US 41 north of Monteagle

SOUTH DAKOTA
BETHLEHEM CAVE: off US 14, 20 miles northwest of
    Rapid City
JEWELL CAVE NATL. MONUMENT: US 16, 15 miles
    west of Custer
NAMELESS CAVE: SR 40 west of Rapid City
RUSHMORE CAVE: US 16A, 5 miles east of Keystone
SITTING BULL CRYSTAL CAVERNS: US 16, 10 miles
    south of Rapid City (This cave has a phenominal dis-
    play of crystals.)
STAGE BARN CRYSTAL CAVE: 11 miles north of Rapid
    City off I-90
WILDCAT CAVE: SR 40 between Rapid City and Silver
    City
WIND CAVE NATL. PARK: US 385 north of Hot Springs
    (Look for the great boxwork formations prevalent in
    this cave.)
WONDERLAND CAVE: I 90 or SR 385 near Nemo

TEXAS
CASCADE CAVERNS: I 10 northwest of San Antonio
CAVERNS OF SONORA: off US 290 in Sonora
CENTURY CAVERNS: SR 474 I0 miles from US 87 near
    Boerne
COBB CAVERNS: SR 195 I0 miles north of Georgetown
NATURAL BRIDGE CAVERNS: SR 281 north from San
    Antonio to SR 1863
TEXAS LONGHORN CAVERN: US 281 south of Burnet
WONDER CAVE: I 35 in San Marcos

UTAH
TIMPANOGOS CAVE: SR 80 near American Fork
A group of unnamed caves is located off US 89 north of
    Kanab

VIRGINIA
BATTLEFIELD CRYSTAL CAVERNS: US 11 near Strasburg
DIXIE CAVERNS: US 11 or I 81 south of Roanoke
ENDLESS CAVERNS: US 11 south of New Market
GRAND CAVERNS: US 340 in Grottoes
LURAY CAVERNS: US 211 in Luray (perhaps one of the most impressive caves in the East)
MASSANUTTEN CAVERNS: off SR 620 near Keezletown
MELROSE CAVERNS: US 11 10 miles south of New Market
SHENANDOAH CAVERNS: I 81 and US 11 north of New Market
SKYLINE CAVERNS: US 340 near Front Royal

Giant's Hall, Luray Caverns.

Discovered in 1878, Luray Caverns is world famous for variety and profusion of formation as well as a wide range of natural color. Giant's Hall features the Double Column, one of America's most spectacular cave formations.
(photo courtesy of Luray Caverns)

Totem Pole, Luray Caverns.

It took the Totem Pole at least 1 million years for the stalactites from the ceiling to meet the stalagmites from the floor of the limestone caves.
(photo courtesy of Luray Caverns)

Titania's Veil, Luray Caverns.

One million years old already, Titania's Veil is still building at the very slow rate of one cubic inch in 120 years. The white is pure calcite, while the darker areas have been colored by iron oxide.
(photo courtesy of Luray Caverns)

WASHINGTON
(L) JARRELL CAVE: Cat Inlet in Vaughn

WEST VIRGINIA
(NOTE: There are a great number of caves in W. Virginia
    too many to fully note here. Contact local grottoes or
    the State Geological Survey.)
ORGAN CAVE: SR 63 near Ronceverte
SENECA CAVERNS: US 33 in Riverton
SMOKE HOLE CAVERN: SR 4 west of Petersburg

WISCONSIN
CAVE OF THE MOUNDS: US 18 and 151 west of Mount
    Horeb
CRYSTAL CAVE: SR 29 west of Spring Valley
EAGLE CAVE: off SR 60 northwest of Muscoda
KICKAPOO CAVERNS: SR 60 15 miles southeast Prairie
    du Chien
LOST RIVER CAVE: US 18 west of Mount Horeb

WYOMING
(W) DARTON'S CAVE (and others in Black Hills Region)
    off US 14, near Sundance, in Crook Co.
(W) HORSETHIEF-BIGHORN CAVES from Lovell in Big-
    horn Co. to Montana border
(W) near Bosler, in Albany Co. on or off US 30-287,
    GRAND VEDAUWOO CAVERNS, HORNED OWL
    CAVE, and many other small caves
(W) near Jackson, in Teton Co., on or off US 26-187, nu-
    merous small caves

As in all things, it is interesting to note that man has a
relatively limited imagination when it comes to naming
caves. If you run your finger down this list, you will find a
variety of Crystal, Lost River, Lost, and Endless caves or
caverns. Maybe that is how they seemed to the first set-
tlers, but those who selected Indian names did a service
to amateur anthropologists by continuing some of the my-
thology that inevitably surrounded caves.

Chapter 5

Going Down in Caves

"Let's see — equipment is all set, I'm in shape and I'm ready to go. . .. Good Lord, will you look at that entrance? There's no bottom!"

That could be you peering into any of a number of American caves. You're ready to do some serious caving, but how do you go about it? We previously mentioned some of the aspects of caving — crawling and walking, climbing and descending. The first two are self-explanatory — either you walk down the passage or you crawl. That is horizontal caving. But many techniques are required for safe and effective travel within a cave. Some knowledge of rock climbing techniques is helpful, but caving has developed a style of its own, and we call it *vertical caving*. While this chapter is not meant to be a substitute for actual practice, it does serve as a guideline for subsequent lessons.

Part of the definition of a cave should read. "It goes down." While this is not a hard and fast rule, most American caves feature some sort of downward pitch that requires the utilization of some descent technique.

Jumping and Sliding

In a word — *don't*. There are too many ifs when you engage one of these forms of descent. That solid floor might actually be a thin ledge; optical illusion may make a slope look flat. Trash atop nothing may look solid, but could actually be nothing more than a false floor. That nice smooth mud slope could pan out over a cliff.

Jumping and sliding can cause a caver to lose control over his subsequent actions. Usually, the penalty is an assortment of bruises, cuts and scrapes, and Murphy's Law dictates that you — and not the other guy — could be the lucky one to break a leg on the first day of a promising expedition. Avoid jumping at all costs.

Ladders

Before man was enlightened to the wonders of the advanced cave-descent technique, he had to depend on ladders to lower him into the earth. Now ladders are bulky creatures by design, and temperamental by nature. They twist and buck like wild horses, so, thankfully, most cavers eschew their use today. If you come upon a situation where a ladder is used, especially if it is hanging free, the heels — not the toes — should be inserted into the rungs. This adds stability and prevents twisting, by placing your center of gravity (you sit on it) closer to the ladder, instead of flopping in the breeze.

Chimneying

As well as being a method of ascent and descent, *chimneying* can be the easiest way to get around in a cave. Basically, you can chimney if you can wedge yourself in place against two walls by applying counter pressure with hands, feet and backside. It will permit you to traverse slots that narrow at the bottom, have running water or present a better route if you could just get a little higher up the passage. Chimneying routes are many and should be taken advantage of at the earliest possible moment.

The basic principle of chimneying is counterforce — push against both sides of a passage and you stay up. You accomplish this by utilizing any and all available holds — be they footholds, handholds or wedgeholds. To traverse a chimney horizontally, you can plant a hand and foot on each side of the passage and be off on your merry way. Going up or down, you can press your feet against one wall and jam your back against the other, and you move in inchworm fashion. Your size and diameter rather than any description, will dictate more about how you can chimney. Just be careful about what you chimney against — some surfaces are not quite as solid as they seem. Examine the surface carefully by tapping it with a hammer — better yet, talk to someone who has been there before.

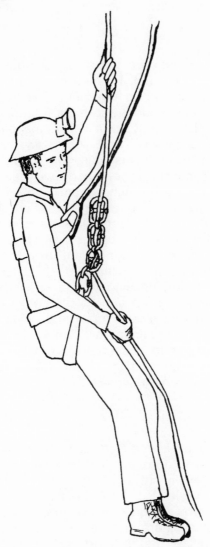

Free Rappel: Here the caver is descending into a well that does not permit him to "toe" the wall as he descends. Note the lower hand controlling the speed of the descent and the upper hand guiding the rope into the carabiner-brake bar rig. Also note the belay line attached to the chest.

Climbing

In some caves, the descents are not so precipitous as to require any form of free descent. Rather, you gird yourself with a *belay* rope and begin to descend. This really is more difficult than it sounds, since the natural tendency of your body is to find the lowest point within the cave — immediately. It's called gravity, and you really have to scramble and claw in order to maintain your grip on the wall. And while every muscle in your body will demand a rest, you cannot quit until you reach bottom.

As you move downward, you change position as the slope becomes steeper, moving from all fours to a three-point position. You also begin to face the wall of the cave, depending on it as a means of support and arrest. But be careful because too much inward leaning will cause your feet to slip out from under, causing you to slide *Slippery or muddy slopes will require an upright position, with feet set against a possible downward trip.* Hand holds of all sorts are very important, requiring the climber to use all sorts of ingenuity in order to conserve his rapidly ebbing strength. Unsubstantial rock and brush should be avoided at all cost. It is simple common sense.

Rappelling

There are times when all is reduced from the sublime to the profane: What do you do about that one drop into that smooth deep well? Well, in days gone by you would just lean back and say "I wish. . .." Now, thanks to an inventive group of cavers, you tie on a belay line, clip a main line into a group of carabiners or a rappel rack and drop off into oblivion. Simplified, that is the rappel.

Originally, the rappel was the tool of mountain climbers looking for a quicker way off the peak; you have seen the films of those long, graceful leaps with the climber bouncing gaily down the cliff. Subterranean rappelling, however, is much less graceful and requires more caution. After all, you cannot see where you are bouncing to.

Rappelling is hazardous and allows little margin for error. Nevertheless, it happens to be the best means of rapid, simple descent available to cavers today. Taken with caution, a rappel will get you where you want to go with a minimum amount of bother, but there are few dependable precautions because the technique depends on two fragile links—the equipment and *you.*

In the early days of caving, rappel work involved wrapping the main line around your body, leaning back into nothing and sliding down the rope. One problem, though, was friction between the rope and the body. Friction means heat and many old timers have rappel scars in unusual places. Thus, pads, either sewn into the coveralls or draped under the rope, were introduced. With modern technique, the rope never touches your body if it can be avoided, and it only plays through your hands as you descend.

However, you should be aware of the old technique, especially for emergency purposes. At the lip of the drop, you straddle the main line facing the anchor. Then you grasp the rope behind you with your right hand, setting it under your hip. Bring the rope around your right side and chest to the left shoulder, and the rope should feed across your back into your right hand as you descend. You can descend more rapidly by releasing some of the tension across the back, and likewise slow yourself by bringing the rope hard against your left shoulder with your right hand.

On Belay

At all times, on all descents, you should be on belay. Simply, belaying is the insurance policy of a descending or climbing caver. A belay line, which is kept taut by another caver known as the "belayer," is a second rope between the caver and a safe anchor point (be it caver or piton) that will prevent the caver from falling an injurious distance if the main line or climbing gear fails. Many a caver has been saved by his belay line when he slipped or fell into space, an environment most men cannot cope with. Belays should be used any time an uncertainty about pitch, depth or ability arises. Never scoff at a person who requests a belay line; he probably has a good reason for wanting one. As a general rule, every time an ascent or descent is made, a belay line should be used.

Perhaps the best belay line available is the Bluewater II
— a non-spin rope that will not snarl or whip about as the

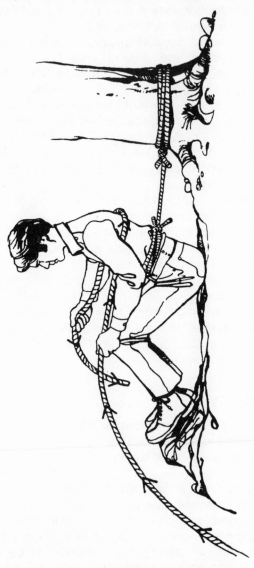

Belaying the Ascent: The belayer is
securely anchored to the tree behind
him. (You can use a rock or piton if
inside the cave.)

spelunker makes his trip. When tying the belay line to your body, the best knot to use is a *bowline-on-a-bight* — a standard, Boy-Scout-issue bowline with two or three additional loops added to spread any impact caused by a belay arrest. (For knot-tying instructions, consult the *Boy Scout Handbook* or *Fieldbook*.) The knot should be culminated with an anchoring *half-hitch* that takes up the remaining free rope and adds an insurance knot. With the knot correctly tied, the caver should be able to endure a jarring fall in free air of about 20 feet without major injury, outside of a few bruises. The additional wrappings around your body will distribute the shock so you do not break any ribs.

Belay signals

During an ascent or descent, communication between the climber and the belayer is essential. In many instances, however, conversation becomes difficult because of distance or natural interference. A system has been developed by climbers (and is accepted through most of North America) to facilitate the transmission of these vital messages. These voice signals are easily heard and understood because of their brevity and phonetic simplicity. A brief scenario follows.

> Climber: Belay on?
> Belayer: Belay on! (if ready; No! if not ready)

Note: Until the climber receives an affirmative reply, he stands still, making no attempt to climb.

> > C: Ready to climb! (Again, this implies a request for confirmation that all is ready.)
> > B: Climb! (If a negative response is given, the climber waits.)

Once the belayer has signalled that he is ready to accept pressure on the belay line through a fall, the climber may either set out on his endeavor or he may choose to test the rope and the belayer:

> > C: Testing! (The climber then either simulates a fall or leans hard on the rope.)

If the climber does not choose to test the belay or has tested and instead begins his climb, he calls out:

C: Climbing!

At this point, the belayer probably will not adjust to the climber's change in position, and there will be slack in the belay line. The climber knows that for every foot of slack in the belay line, there is one more foot he will fall if he slips. Hence, he calls:

C: Up rope!

If the belayer gets over-zealous in keeping the rope taut and starts to pull the climber up the pitch, the man on the rope will shout:

C: Slack!

If the climber wants the rope as tight as possible, he yells:

C: Tension! (not "up slack" or any variation).

Other signals that are self-explanatory include, "Resting!" and "Falling!" When one lets loose with "Falling!" the belayer immediately takes up all the slack in the belay line and braces for a sudden shock. If you are feeding the rope through your left hand around your back, through to the right hand, the best way to lock the line is to grasp the rope firmly with the right hand and pull it across your chest to the left shoulder. This uses your entire body as a friction point and will prevent the rope from slipping.

A final signal used in a climbing situation, and one that can be used by any member of the party, is: "ROCK! ROCK! ROCK!" This signal lets anybody below know that something is falling and the best course of action is to get under cover now!

Remember that falling objects accelerate with distance and even the tiniest pebble can do serious damage to head and health if it drops a few hundred feet.

Avoiding Rope Burns

Currently, several mechanical aids have appeared to reduce the caver's risk of rope-burns. These come in a variety of exotic forms, including the *whale tail* and *rappel rack*, but the most economical and simple of all rappel aids is the *carabiner brake bar rig*, which minimizes the strain on the rope. The rope passes through the carabiners and over the *brake bars* in such a fashion that you may control your descent merely by locking the rope with a quick motion of the hand. The other two aids work in a similar fashion.

The technique used in rappels with mechanical rigs differs in form, but not in theory, from the body rappel. The rig replaces your body as the hot spot, by connecting the rig to your seat harness after tying in your belay. Then, feed the rope through the rig, over and under the bars. In essence, the rig has become part of your body, and the descent is performed similarly to that of the body rappel, with the rope feeding through your right hand into the rig and out through your left hand. Speed control is maintained by changing the angle of the rope feed into the rig with your right hand. You can even stop midstream by looping the incoming line over the rig in a sort of overhand knot.

Despite the benefits, the carabiner brake bar method has certain disadvantages. Among them, the rig is susceptible to rope kinks and knots that may cause sudden, unwelcome stops. Also, rapid rappels generate a great deal of heat, which is a definite no-no with dynamic ropes. Many times, the brake bars will get as hot as 250° F., and this temperature will melt a rope, so be cautious with your speed. Yet, all in all, this rig is excellent for drops up to 200 feet, and that will be the maximum for most of your early efforts.

Other types of mechanical rigs generally are used for much greater descents when friction toward the bottom end of the rope is more important. Near the top of the descent, with 40 or so pounds of rope winding down into the depths, the friction in any rig is great. But, near the bottom, many rappellers have been unpleasantly surprised when their friction disappeared, along with the weight of the rope, and the rappel rack was developed with this problem in mind. Near the top, the rope is fed over a few brake bars, and as friction is reduced, the caver merely clips a few more brake bars into the system, with the rope passing over six or more brake bars. This unit has been adopted with great success in descents of 1,000-plus feet. Whale tails work on a similar principle, but they dissipate the heat more efficiently because there is more metal to radiate heat.

Yet, despite the friction, many cavers persist in zinging

down the rope at high speeds. This causes wear on
equipment and fellow cavers' nerves, in addition to running the risk of melting the rope or pulling loose the anchors as the fast-descending caver slows near the bottom.
Then — oops!

How You Go Down

Even with all the rappel rigs, you will discover that there is
a certain trick to descending on a standing rope. A long
time ago, the line of thought was that you had to lean
back at a sharp angle, playing your feet along the wall of
the drop. This position is rather uncomfortable as it requires you to maintain a rigid posture. Better that you
should descend in a standing or sitting position, toeing the
wall with your feet below you to touch down without a jarring thump. This laid-back position also is good for pitches
that are not absolutely vertical, producing better traction
and footing.

Other Types of Rappel-Air

Some cavers have had the bad luck to discover that air is
not an effective medium for rappel descents. Usually this
occurs when that 120-foot rope dropped into that 100-foot
well turns out to be a 120-foot rope dropped into a 130-
foot well. That last 10 feet can be a killer, and to avoid
such problems, a wise precaution would be to tie a large
knot in the end of your main line so that you will stop
when you reach the end of your rope.

Setting Up to Go Down (or Up)

Before you go anywhere, you must anchor your ropes in a
fashion that will allow you to distribute the strain on the
standing rope caused by your descent. The most common
anchor is to tie the main line to a tree — without using
knots. This is called *tensionless rigging*.

Although not always feasible, this method is perhaps
the easiest to set up, and the strongest of all anchors.
Basically, the tensionless anchor is made by taking the

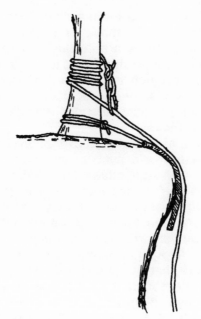

Tensionless rigging.

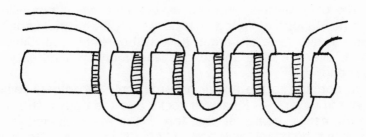

Whale Tail

standing rope and wrapping it three or four times around the nearest tree. The anchor is completed by attaching the short end of the rope to the mainline using a bowline knot. The proper term for the above wrapping is a *bowline-on-a-coil.* The reason knots are avoided in the actual stress areas, the coils around the trees, is that a knot can weaken a rope by as much as 25 percent. Avoid at all costs anchors using knots, or the additional surcharge may be your life. Other possible anchors outside of the cave include rocks and cars.

Sometimes none of the above is available, especially deep within a cave. In such cases, you can drive pitons into the cave wall and attach carabiners and slings to a master carabiner. The rope then is run through the main link. This method, however, is only usable if cracks are present and you can drive solid pitons. Otherwise, a sudden shock may very well pull the pitons free. Also, try not to use pitons to anchor standing ropes. If you must, at least make sure that the pitons are secure. You can determine this by the sound the piton makes while it is being driven. If it goes *ping-ping-ping* in ever-increasing pitch, the piton is safe. If it goes *thunk,* seriously consider leaving before the whole face of the rock slips away.

As previously mentioned, all ropes that cross rock or earth should be protected in some fashion. The lip of a well is an obvious trouble spot and special padding should be used here. Usually a canvas pad, attached to the anchor point to secure it or main line using another rope will do the trick.

There is another problem cavers face when rappelling — the danger of frying a rope as it passes the carbide lamp. A slit piece of garden hose will cure this evil. Simply slip it over the rope and hold it with your left hand as you descend.

In addition to the above hazards, only two others really concern cavers — kids with rocks and wise guys who rip-off ropes. A guard at the top will prevent Junior from dropping boulders down the shaft in an attempt to play nine pins with cavers. A guard armed with a Mexican Speed Wrench will eliminate the jokers who think that removing standing ropes while cavers are below is great fun.

Going Home to Mother

Once you have gone this far, you are faced with the sticky problem of getting out again. Generally, this poses little difficulty — you simply reverse the descent action. Or do you?

Mountaineers really cannot relate to the difficulties cavers face when trying to get out of a hole in the ground. To them, time is not of the essence. Besides, all handholds and cracks are readily visible. Yet a caver, when faced with an ascent through an icy waterfall, does not have the luxury of choice — he has to get out quickly. Thus, enterprising spelunkers have evolved techniques and mechanisms to facilitate rapid ascent and, since ascents normally are made when the caver is exhausted, the uphill trip must be made simply and efficiently, as well.

Rock Climbing

Before the final ascent out of the cave, Joe Caver, often must surmount faces, pitches and cliffs within the cave. This calls for the technique known as rock climbing. In its most elemental form, rock climbing involves the process of going up cliffs and other surfaces without the aid of stairs or ladders. The climber takes advantage of all hand and toe holds, but reserves his legs for the actual climbing; he does not pull himself up with his arms. First, his arms are not strong enough to pull his tired body very far. Second, suppose he does pull himself up with his arms, but does not find a place for his feet? What then?

There are two cardinal rules in rock climbing. One: pace yourself to maintain a constant momentum — even if it is slow. Two: keep a relaxed, upright posture. If you lean in, you run the risk of thrusting your feet right out from under you.

A full discussion of rock climbing technique is beyond the scope of this book. However, some common sense methods suggest themselves: Never overextend yourself. Upward steps should never be longer than a foot or so. Make short movements rather than large ones; they require less effort and leave broader margin for error. Use your hands for balance, not for pulling power. Finally, patience is your watchword. Do not hurry anywhere or your next stop might be disastrous.

Rope Ladders

Cavers in the United States never really experience the joys of swinging to and fro on a rope or cable ladder. Europeans, on the other hand, use them to conquer huge drops of hundreds of feet. In the U.S., ladders flourished briefly but soon were replaced with rope and rock climbing. The main problem with rope ladders at depths of more than 60 to 70 feet is that they tend to whip around and develop the most unruly tangles when a climber is on them. Yet, on shorter pitches, ladders can help move a large number of cavers in a short period of time, especially if the party is blessed with a number of novices who are unfamiliar with standing rope techniques.

Like ladder descents, the key to effective ascents is found in keeping your body's center of gravity as upright and as close to the ladder as possible. Thus, put your heels into the rungs and climb around the ladder. If you climb a rope ladder as if you were on a rigid ladder outside your home, you will end up leaning farther and farther backwards as your feet push the ladder away from you.

Again, as with all other descent techniques and climbing methods, wear some sort of harness, and always be on belay. Too many problems can crop up in a short climb to work any other way.

Ascent Knots and Mechanical Climbing Devices

While ladders offer a reasonably safe way to handle short pitches, other methods have been developed to conquer those wells of 100 feet or more. Earlier, rappel techniques were discussed in terms of getting into the hole. Now, its time to look at the use of the standing rope in ascent.

Think about this for a moment: You are at the bottom of a deep well with totally smooth, water worn walls. Dinner is on the stove back home, and you do not have the time to set pitons. What do you do? Well, you could grasp the rope firmly in both hands and shinny up, but that is probably the one sure way to prolong, and perhaps make permanent, your stay in the cave. More novices have been hurt or killed by this than any other single foolhardy act. The reason is best explained by this question. How can you rest 75 feet up the rope if you have to hang on for dear life to stay in that position? Admittedly, there are some situations where unattached rope use is acceptable — i.e., in a tight chimney where you could use a rope as a third suspension point. However, this in itself is not climbing, and should never be construed as such. Like skin diving in caves, free rope climbing is a killer.

The logical answer to the free climbing question is some sort of aid that will hold the caver in place on the rope without requiring any physical effort from him. Ascent knots and mechanical climbing devices working on the friction principle have been developed to fulfill this need.

Unfortunately, discussions on knots and mechanical equipment invariably sound as complex as lectures given by a NASA engineer. The terminology involved applies to caving alone and is not known generally outside spelunking circles. *Prusik, helical, Hedden* and *Bachmann* knots are among the more common nonmechanical aids in use today. For the novice, however, perhaps the Prusik knot is the best and easiest to use.

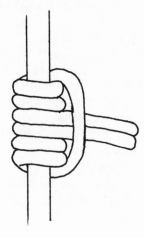

Prusik knot.

The Prusik knot works on a very simple principle — friction. With the knot, you are able to freely slide a sling rope up a standing rope, but when weight is applied in a downward direction, the knot firmly grips the rope. Prusik barely fits the definition of a knot, since it is merely a series of loops around the standing rope (see diagram).

A Prusik knot is formed by passing a loop of the sling rope around the main line two or three times, each time going through the eye formed by the sling. A small space is left between each coil. Usually, two passes suffice, making a four-coil knot, but on muddy ropes, a six-coil Prusik works better — giving more friction and allowing greater gripping power. The Prusik allows for a greater margin of error than most knots and consequently, it climbs well under the most incredible circumstances. The best demonstration of the knot's dependability is that it is used successfully in series on oil-coated pipes.

Of all the mechanical ascenders on the market today, two — the Gibbs and the Jumar — probably are the best for cavers, being easy to use and service. The Gibbs, for practical purposes, is a hand-held, mechanical Prusik. In fact, its use is identical to that of the Prusik. The Jumar, on the other hand, grips the rope with a toothed, eccentric cog. One control on the device governs the cam — releasing it so that it can be slid upward. Downward pressure on the device sets the cog into the rope. The other control permits the gate that guides the rope to be released and allows the unit to be placed on the rope or removed. Do not confuse the controls. The slings for chest and feet are attached to the ascenders through an eyelet provided.

One problem with the Jumar is that it has a number of moving parts, rivets and pins. The old rule of thumb still holds — the more parts, the greater the chance of mechanical failure. While the Jumar is a valuable device, its cog teeth may dull and need replacement; pins may pop out; rivets might break. And the topper is that the Jumar is expensive — almost $65 a pair. Drop one into a pit and mourn — a lot of cash can go down the tube in an instant.

Climbing with a Prusik or a Jumar is not as complicated as it may seem, though, the first trips may seem endless,

A: *The climber is in a full upright position with his weight firmly distributed on both lower leg (a) and the upper leg (b). (Note the chest harness is attached to the lower leg sling.)*

B: *The climber swings the upper leg out in front of him and slides the upper ascender up the rope until all slack has been removed from the sling. He then supports the weight of his body on the upper leg in a squatting position.*

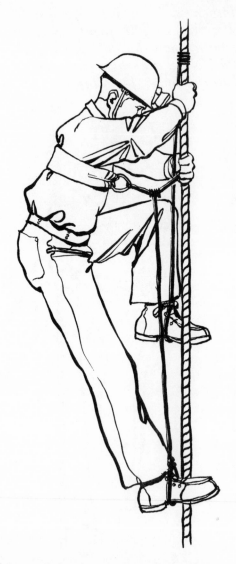

C: Climber assumes semi-squatting position, supporting his weight on upper leg while sliding the lower ascender up the rope until he is able to support his weight equally on both legs. Then, he stands up and assumes position A.

even if you are in top physical condition. It takes a lot of practice but, eventually, you will be able to scoot up a rope in no time at all.

Earlier, we mentioned chest and Swiss seat harnesses. Again, these come into play. But, you still need another piece of equipment for the ascent — foot slings. Two are necessary — one for each foot. Take two lengths of ⅜-inch hemp rope, one about 11 feet long and another about nine feet long. Splice the ends of each length together to make two loops — one five feet long and one four feet long. Make two loops at one end of each sling for your feet, and you are in business. When you are set up on the rope, attach the chest sling (another loop) to the long leg rope, and the Swiss seat to the short leg rope (using a length of line between one and two feet, depending on height and girth). In this fashion, you are attached to the rope in a vertical position, keeping your center of gravity close to the line. If you do tire as you climb, you can squat down until the seat harness line is taut and your weight is supported by the Swiss seat.

The climbing motion used when the Prusik is involved can be rhythmic and graceful. To start a climb with Prusiks, raise the leg on the short sling until you can slide its knot up to the long sling knot. Then, supporting your weight on the short leg, slide the long leg knot up the rope as far as you can reach, swinging the long leg out in front of you as you do so. Then support your weight on the long leg and begin the process again.

You should be able to gain about two to three feet for every cycle of Prusiking, but if you tire, shorten the distances and move more slowly. Some cavers, by the way, will use three knots — two for the legs and one for the chest — but that requires one extra unnecessary motion. No matter what is written, no account of Prusiking will provide an adequate guide. Make an outfit and try it yourself. Then, decide the best technique for you and go with it.

The Gibbs Ascender

While the Gibbs Ascender is a simpler device than the Jumar, its use tends to be more complicated. Called a "rope-walking" ascender, no hand action is required to operate the Gibbs. This unit also operates on the cam/eccentric gear technique, gripping the rope with a toothed cog. The Gibbs, however, does not have a locking mechanism that must be keyed out of the way before the ascender can be moved. The slightest upward motion will release the unit and relatively little weight will cause it to lock onto the rope.

When setting up your Gibbs climbing outfit, you need three ascenders in order to "walk" the rope properly. One is strapped to the inner side of the lower foot, usually at ankle level. Another is attached to a shoulder harness rigger somewhat like a Sam Browne belt to your Swiss seat (see diagram). A third, called a "floating Gibbs" is operated by the upper foot, but attached to the shoulder harness with a length of elastic shock (stretch) cord. This Gibbs is activated as you raise the upper foot. Upon release, the ascender is pulled upward by the elastic cord and locked in to place when you apply weight to the upper foot.

A word of warning about using Gibbs Ascenders: Since you are not using your hands as control factors (something you will have to get used to — it's a natural human inclination to grasp at anything when under stress), you must keep your center of gravity as close to the rope as possible. Hanging sideways on a rope is no excuse for being slow. If you have a pack or another load, strap it to a rope from your legs or hips. Then the weight will be distributed in a vertical direction below you.

Another Word of Warning

All mechanical ascenders, like any other piece of equipment, must be examined regularly for wear and tear. Cams, rivets and pins must be checked as they are the vital parts of the unit. If you harbor suspicion about any part of an ascender, or if the whole unit has been acting up, replace it at once — diecast and punch-pressed parts are much less expensive in both dollars and trouble than broken bones. Also, before each excursion, examine the slings on each ascender or Prusik for abrasion or cuts.

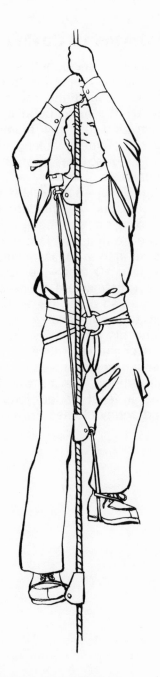

Gibbs Rope Walking Technique.

The Ups and Downs of Caving Life

Invariably, when cavers get together, yarns, anecdotes and lies (the mix varies in proportion to the age of the story) fly thick and fast around the room. Climbing stories, however, seem to be the most popular and most drama-laden. It is truly amazing just how many cavers have had problems with their ascenders locking into place, slipping on the line or working so well that the climber literally zinged out of the cave.

Our favorite story involves a certain young caver (who will remain nameless) and an incident involving the free rappel. It seems that he was equipped with a fifty-foot rope (he did not realize that 100 feet is considered a minimum) as he peered into a passageway in a New York cave. The avenue was 10 feet wide, and the bottom was 45 feet below the notch where he was stationed. Since the notch was a one-man affair, he anchored the rope to a fellow caver in an adjacent passage and set up his belay line. Then he began his rappel descent. The first 50 feet were not bad; it was the last seven that were annoying, especially the dunking in three feet of cold water. The insult added to injured pride was the wait for his partners to toss a line that was at least 57 feet long. The moral of the story: Wet feet and soggy clothes are no fun.

Chapter 6

Caving May Be Hazardous to Your Health, but It Doesn't Have To Be

Here and there throughout this book have been gentle reminders and subtle cautions about the inherent dangers of caving. Now, it is two-by-four time, so be prepared to be hit over the head with some obviously simple points concerning safety while caving. All outdoor sports are dangerous to some degree. Sailing, rock climbing, skiing and even hiking have their dangers and, to many, these dangers — recognized and prepared for — serve to make the sport more appealing. But while something may hold the potential for danger, there is no reason why it cannot be controlled. Caving is no exception, and the best way to ensure that you come out of a cave the same way you went in is to avoid stupid mistakes.

Stupidity is *the* cave killer, followed closely by its cousin, carelessness. In the case of the novice caver, stupidity manifests itself through seven major categories:

1. Caving Alone. If you want to cave alone, be sure to include a tombstone in your pack. There is no way a single caver can be aware of everything that is happening in a cave. Falling rocks, firm-looking, but actually soft ledges, and quick drops pop out of nowhere. Even a twisted ankle can prove fatal to a caver who has left his group to go it alone.

2. Ignoring Outside Weather Conditions. You may think you are safely inside the earth, far from the outside world, but what is outside eventually finds its way inside. If rainstorms are predicted, you can be sure that the rainwater will find the path that took you into the cave, and what was originally dry and solid may quickly turn into a raging stream or waterfall. Flash flooding also is a problem, not only because of the moisture, but because entire caverns have been known to fill up in seconds, as well. Thunderstorms are even worse, since the electrical discharge of a lightning bolt can easily find its way into the depths of a cave.

3. *Improper Clothing.* The ambient temperature of a cave is generally moderate, but there is a great difference between 50° F. inside and 50° F. outside. First, there is no sunshine to warm you. Second, the length of time spent in a cool cave has cumulative effects. If you do not dress warmly at the start, there is no way you can avoid becoming very cold later on, and you have set the stage for a grand attack of hypothermia.

4. *Drinking Alcohol.* While a good snort can give you a nice warm feeling, don't fall victim to the St. Bernard syndrome: Those cartoons of man's best friend with the keg of brandy rescuing the fallen skier are pure poison for cavers. Alcohol lowers the temperature of the body, and you are supposed to stay warm, remember?

5. *Free Climbing.* Never, repeat, *never* climb under any but the most controlled circumstances. "Look before you leap" is the first commandment of caving. "Hold on tight" is the second.

6. *Clowning.* The purpose of spelunking is enjoyment, and no one wants to deny a caver his fun. But a cave is no place for practical jokes. Moving a rope, playing hide and seek, running, jumping and generally horsing around not only make enemies, they also make corpses.

7. *Ignoring Expert Advice.* Your leader knows more about caving than you do. Trust him. If you doubt his advice, ask him to explain his reasoning, but never reject it out of hand. His is the voice of experience, and all that.

If you can accept these seven points graciously, you should have no major problems caving, but cave safety involves more than this. It actually begins before you cross the entrance of a cavern. The first thing to do before you set out is to notify someone in the outside world that you are going caving. Give the name and location of the cave, pinpointing the specific areas to be explored if it is a relatively large system. Also, let him know when you plan to return. In this way, he can notify the authorities if you are unreasonably delayed.

Make sure your group leader is aware of any physical limitations you may have. It is expected that members of the group will have varying degrees of expertise, and the abilities of each member must be known and evaluated before entering the cave to be sure no attempts are made to exceed those limitations. Climbing and crawling may appear to be natural activities, but they are not. Each member of the group should be taught and tested on the various techniques of cave exploration before beginning. A quick review may seem superfluous, but it can prove vital in determining the limitations of the group. And, in many cases, any deficiencies can be corrected before the trip is begun.

The final caution before beginning a cave exploration: Check your equipment. While a frayed rope may be strong enough for one more trip, there is no need to press your luck. On the same note, too much carbide is better than too little, and extra weight of a complete change of clothes is negligible when compared to the effects of hypothermia caused by being soaked through with water. Once inside a cave, all you have is your equipment. Do all you can to make sure it will not fail you.

Blisters

Regardless of safety precautions, mishaps do happen, and a variety of injuries can and will be sustained underground. The most prevalent cave malady is the blister, caused by too much walking in loose shoes. The best thing to do is to relieve the pressure of the blister by puncturing it near its base with a sterilized needle. Rubbing alcohol is best for sterilization, since it is not wise to light an open flame in a cave. After the blister is drained, it should be covered with a pad to prevent irritation. If a blister is already opened before you can do your handiwork, clean the area with an antiseptic to prevent infection.

Cuts, Scrapes and Bruises

Next to blisters, cuts, scrapes and bruises are the most common injuries. Most cuts are minor, and these will require only simple cleaning and bandaging. Soap and water are the best cleaning agents, but highly impractical in a cave. Instead, most first aid kits contain medicated towelettes that will clean the injury well enough to prevent infection. Deeper cuts usually cause bleeding, and this can be stopped by applying heavy pressure with a tight bandage. After the bleeding is under control, the wound should be cleansed thoroughly and a fresh bandage applied.

Scrapes are more annoying than they are dangerous. Quite often, they are more painful than severe cuts, but anyone who has skinned his knee knows that the pain is only temporary. Since scrapes do tend to bleed slightly even after they have been treated, it is best to use a non-sticking gauze pad to cover them. The pad should be applied loosely, so outside air dries the wound and a protective scab can form.

Bruises, too, are often more annoying than dangerous, and the primary problem here is swelling. The best thing to do for swelling is to apply cold, moist packs to the affected area. After the initial cold, hot packs may be used on succeeding days to increase the circulation around the affected area and promote healing of the damaged tissue.

Sprains and Strains

Because the terrain in caves generally is rough, sprained ankles are a commonplace occurrence. Sprains, of course, also can occur to other parts of the body. Regardless of the site of the injury, the treatment is essentially the same. Technically, a sprain is the tearing of ligaments around a joint, and the first thing to do is to apply moist, cold packs to reduce swelling. After the swelling has gone down, an elastic bandage should be applied in a figure-eight pattern above and below the joint for support. Again, heat on succeeding days will speed healing.

A muscle strain, while similar in effect to a sprain, is fundamentally different and not quite as serious. In a strain, the actual muscle is torn and, again, cold should be applied to reduce the swelling. Like the sprain, the tried-and-true elastic bandage should be used afterward, wrapped tightly around the affected area to support the torn muscle.

Fractures

Sprains and strains are relatively minor injuries and, in most cases, after the initial treatment, the caver should be able to go on his merry way with little difficulty. Fractures, however, are much more serious, and there is little a caver can do for a suspected fracture victim besides rudimentary first aid treatment — and going for help.

The best thing to do for a fracture is to splint it. Since suitable splinting materials will be scarce underground, you will have to use your imagination. Arm fractures can be bound to the torso, and leg fractures to the other leg. Again, these are only stop-gap measures and, unless you are expert in first aid treatment, you should just try to keep the victim still until the rescue team arrives.

Shock

The same is true for other injuries of a serious nature, and your biggest problem underground will be protecting the victim from injury-induced, traumatic shock. Shock can be caused by most severe injuries and is aggravated by a delay in its treatment. If a member of your group has sustained a severe injury, do not try to determine whether he has gone into shock. If the pulse is rapid, the skin cold and flesh pale, treatment for shock is indicated. The degree of shock is affected by the amount of change in body temperature and, in a cool cave, the chances are the change will be severe.

Keep the victim lying down to improve blood circulation throughout the body. In some cases, his condition may be helped by propping up the feet. Insulate him from the cold floor with spare clothes and cover his body just enough to maintain a constant body temperature. If there are no apparent broken bones, you can aid his circulation by massaging his limbs.

Hypothermia

Next to shock, the most dangerous condition one can encounter in a cave is hypothermia. As noted previously, hypothermia results when the body loses more heat than it generates. In caves, it usually occurs after the victim's clothing is soaked through. The best cure is prevention— if you fall into a pool or stream, or you are otherwise soaked, get out of the wet clothes as quickly as possible. If you don't, and hypothermia begins to set in, you will begin to shiver from the chill. As your body temperature begins to drop, the shivering will stop and your muscles will become stiff. You also will begin to lose touch with reality, eventually entering a stupor-like state before losing consciousness.

If you can somehow avoid the final stage, and you still have not gotten out of your wet clothes, now is the time to do it. Next, drink hot liquids to heat your insides; the hotter the better. This will warm your body where it needs it most — the core organs. External heating also helps, and the stomach, neck and face are the best entry points. After body temperature begins to return to normal, return to civilization as quickly as possible and see a doctor immediately.

The American Red Cross publishes an excellent first aid manual in much greater detail than can be included here. All spelunkers should carry a copy of it with their first aid kits. It is an even better idea to take a certified first aid course before you attempt any extensive caving. Never assume someone else in your group is a first aid expert. It may turn out to be true, but it's not worth the risk to let that responsibility fall on someone else's shoulders, especially if that someone else is the victim of an accident.

Chapter 7

Cave Rescue

In the world of caves, there is an unwritten rule that what goes down must come up. In other words, every caver who ventures into the depths of the earth is responsible for his own reentry into the world of surface dwellers. Every rule *does* have its exceptions, of course, and there are always those who subscribe to the view that rules were made to be broken. Hence, the birth of cave rescue.

The process whereby luckless spelunkers are extricated from the bowels of the earth — in a manner not unlike the operations of modern mountain rescue units and other emergency teams — has happily improved as steadily as the number of underground adventurers has increased over the years. Thanks to the unified efforts of the National Speleological Society and its offspring, the grottos, the disorganized, sometimes fatally inefficient rescues of yesterday are, for the most part, mere shadows of the past. As we all know, however, skeletons sometimes escape from their closets, and it is the duty of every caver — beginner and expert alike — to relegate old mistakes to their rightful place — gone forever.

The first and foremost principle of cave rescue is *prevention:* The best way to handle cave rescue is to avoid the need for it. Be prepared to come out of a cave on the same power that took you in. If you follow the advice in the preceding chapters, you'll be one step ahead of the hodag that lurks in the shadows of every cave. And if you realize that even the most experienced spelunker can get into trouble, you'll retain a sense of humility that could save your life. The dread hodag sometimes surprises even the most thoroughly prepared spelunker, so it doesn't hurt to have an ace up your sleeve.

Years of practice may be needed before the caver is proficient enough to extricate himself or another from any sticky spot. And most cave rescues require more than one or two participants to succeed. So the smart caver takes

out insurance against accidents, devoting a little extra
time in preparation for the emergency that, hopefully, will
not occur.

Today, most cavers are well informed, and as time goes
on, the number of reported cave accidents decreases.
But, whenever cavers get together, harrowing tales are
likely to fly around the room. Just about any seasoned
caver will need little encouragement to relate one or more
tragic tale of spelunking mishaps, and, like the fabled fish
story, these tales tend to expand with each telling. Polite
listeners never question the factuality of a story, automati-
cally assuming that it is based on history plus imagination,
and that the moral of each fable is what is important.

The following are some of the caver's legends that
we've come across in our travels. They are not meant to
frighten you away from the sport; nor are they meant to
be patronizing. They are included merely to illustrate to
beginners exactly why simple steps in preparation are so
vital to their safety and enjoyment in caving.

——John C., a summer counselor at a boy's camp, had
made big plans for his day off. Three of his friends were
to meet him at the camp, and the foursome would drive to
the site of a nearby cavern system. The group was well
prepared for its caving expedition and John had left word
of their destination with the camp director.

While John loaded his gear into the car, the excited
discussion turned to another cave site about 25 miles farther away. Swept up by intriguing images of pioneering
some little-explored caves, the four agreed to switch their
destination and drove off on their adventure, expecting to
return by nightfall.

Twelve hours later, a 50-man rescue team was scouring
the caverns named by John's employer when he reported
the cavers missing. Meanwhile, the hapless adventurers
were crawling hopelessly in circles through the intricate
passages of a cave 25 miles away — lost, wet, cold and
exhausted. If it had not been for the quick thinking and
diligence of the search director, the party still might be
missing. A native of the area and an expert caver, the director recalled a similar party of lost cavers from his
youth. Taking a life-saving gamble, he sent a squad of
skilled spelunkers to the site of the other cave, hoping the

missing cavers actually had changed their plans at the last minute.

They were found in a labyrinthian area not far from the entrance of the cave, where they had ceased their circuitous trek and had all but given up hope of being found. No one was injured in the search, but scores of individuals had been led on a wild goose chase which could have ended in tragedy. Despite their hours of preparation, John and his friends had failed to take the most basic of all precautions — letting those above ground know exactly where they were going. No matter how silly it seems, taking the time to stop and phone in your change of plans could save your life. The veteran caver, of course, is likely to seek out virgin territory, but he prudently will leave directions to the specific area he plans to explore.

Mary H. and her brother, Bill, were avid spelunkers. In the spirit of true cave enthusiasts, they had been teaching several friends the techniques of caving, had helped outfit them, and had taught them the rules of first aid. Mary and Bill were skilled cavers, but previous companions had included only equally experienced cavers. Planning their first trip with their neophyte friends, they chose a fairly simple cave close to their midwest home.

Spirits were high as the group set out, and the first stretch of the adventure went smoothly. As luck would have it, though, one of the eager beginners slipped, twisting and lodging her ankle in a narrow crevasse. Try as they might, Mary and Bill were unable to free the victim's swollen foot, and they decided to phone for help. The self-appointed messengers expected to exit from the cave with little hassle, drive to a phone to call for a rescue unit, and have their friend extricated straight away. They assured their companions that there would be no problem and that they would return posthaste.

An hour or so after the pair had set out, those left behind began to get nervous: Perhaps Bill and Mary never got out of the cave. . .maybe they were having car trouble . . . maybe they couldn't find a phone. Tension in the cave built as these panicky thoughts raced through their minds, and the novice cavers decided there must be something they could do, especially if Bill and Mary needed help. Finally, two of them set off in the direction they thought their leaders had taken, leaving the victim and a lone companion behind.

Bill and Mary returned to the site of the accident with two expert cavers and the appropriate rescue equipment — food, medical supplies, blankets, stretchers, pulleys and other gear. Rescue of the injured caver turned out to be pretty routine, but what of the two cavers who had gone after Bill and Mary? To make a long story short, the headstrong beginners had lost their sense of direction and wandered deeper and deeper into the cave, rather than back to its entrance. It took the rescue team three times longer to locate the wanderers than it had to extricate the original victim. Again, the ending was a relatively happy one, but the price — in cost, time and human effort—was much too high.

— Sam G. had been exploring caves for five years. He'd seen ice caves and limestone caves, large caves and small caves, simple caves and the truly challenging caves. Fortunately, he felt, he'd never encountered the need for a rescue. One fateful summer day, Sam and three other caving enthusiasts headed for a cave known for its complexity. The group had been working up to this challenge for a year, and were confident that they were ready for it. All the preparations had been made with extra care, all equipment had been double-checked, maps of the charted passages had been studied again and again.

Sam and his companions did everything right. They took care when they reached portions of the cave that didn't appear on their maps, proceeding with the caution and know-how of experts. But, sometimes all the preparation in the world can't help, and one of Sam's companions slipped into a 10-foot pit, fracturing his right arm. Unable to move or be moved with the equipment they had, the victim was forced to remain in the pit while two of his friends went for help. The messengers reached the entrance of the cave in good time, proceeding to the nearest phone.

Sam, yielding to the creeping panic he felt as they rushed clumsily out of the cave, immediately called the operator to report the emergency. He then returned to the scene of the accident to await the rescue team. He didn't have to wait long. Before the rescue team could arrive, scores of others did — friends and relatives of the cavers, curious onlookers, paramedics, firemen, policemen and any other well-meaning volunteers. The scene became chaotic in a matter of seconds — seconds that were all-important to the victim. The crowds thwarted the initial organization of the rescue team and block with their cars and bicycles the arrival of the appointed director. After much delay and confusion, the rescuers finally reached the victim, who was suffering from shock and hypothermia. The young adventurer did recover, but the town is said to have remembered the horror of the incident for years after it.

   — As in the case of Sam. G. and party, Susan T.'s caving accident was a serious one. Susan was a beginner and an enthusiastic one. She had diligently studied the instructions she received from experts, and had borrowed and purchased enough equipment to take several introductory trips. The cave she was to explore that particular day was the most difficult she had ever approached.

   She and her friends were enjoying their trip thoroughly, slithering in and out of narrow crawlways lined with slippery mud, when Susan felt the beginnings of an asthma attack. She had suffered from asthma since childhood, but the attacks were returning less frequently, and she rarely thought about them anymore. Halfway through a tight crawlway, she stopped, gasping from breath. Her companion, ahead and behind, halted their progress, figuring that Susan had merely become temporarily wedged. When they realized what was happening, the rescue operation was set in motion, and the two cavers began the attempt to push, pull and inch Susan to the other end of the crawlway. It was slow, tiring work and, by the time Susan was out of the crawlway, she was too weak to go any farther. The messengers reached the surface in good time and, after calling for help, one of them was appointed temporary rescue director. Meanwhile, those left in the cave busied themselves by keeping Susan warm and quiet, calming her asthma attack.

   The rescue unit, armed with medical supplies, stretchers, and caving gear, was ready to begin its work, when an argument broke out at the mouth of the cave. Apparently, the temporary director was not convinced that the newly arrived expert was qualified to replace him, and a heated discussion raged as the seconds ticked by. Back in the cave, Susan's companions, knowing her condition would worsen if she became nervous, tried to hide their own mounting fears. What was keeping the rescue unit from reaching them? Why was it taking so long? An argument between two so-called "leaders" was the last thing that crossed their minds.

Susan was saved by oxygen carried in by the rescue unit in the nick of time, but her companions decided not to tell her what caused the delay.

Again, there *are* cases in which human error is not the cause of caving mishaps. Sometimes, the poltergeist of the netherworld intervenes, and the unfortunate victims just have to make the best of the situation. In the stories related above, however, the worst of each emergency could have been avoided if the spelunkers had heeded certain all-important rules of cave rescue.

John C.'s mistake was a fairly obvious one, but one that is behind more cave tragedies than most spelunkers would like to admit. Granted, John's situation may have required a search anyway, but if he had had the presence of mind to inform camp administrators of the group's change in plans, the rescue might have been accomplished quickly and easily by a small team of local cavers. This apparently trivial omission brought on the necessity for a full-scale rescue operation — an effort that might have failed if an expert guide had not been available.

Bill and Mary H. made two basic mistakes in handling their emergency. First, they should have taught their companions one of the most important principles of a rescue operation — keeping cool. Those who stay with the accident victim can keep warm and calm if they busy themselves with helpful tasks like heating water or damming streams, emptying ponds and moving rocks for the rescue unit. The siblings' second error was in the fact that they — the only two experienced cavers in the group — both went for help, leaving the tyros to their own devices. If they had prepared their students more thoroughly for emergencies, and left one veteran to lead victim and company, the most difficult leg of the rescue — finding the lost, would-be help-seekers — could have been avoided.

Sam G. made a similar error in judgment. As an experienced caver who'd never encountered any serious trouble, he had become over-confident in his own abilities, forgetting that the ever-present hodag could strike unexpectedly. His companion's accident took him by such surprise that his extensive experience was of no use and he was unprepared to take the necessary steps in a rescue situation. Starting off on the wrong foot, Sam communicated his panic to exactly the wrong recipient. Panic is contagious and is the foremost enemy of an efficient rescue. On the other hand, if Sam had reacted to the emergency with common sense and had been prepared to record and transmit vital details to the proper source, his sense of command probably would have spread to the others involved, improving rather than hurting the situation.

Two basic errors were made by Susan T.'s caving party. First of all, the caver always should take into account any physical disability that could hinder progress underground — and should inform caving companions of that disability so that they will be prepared to handle the situation. If Susan had carried the medication that she thought she wouldn't need, the rescue might have been unnecessary.

This rescue was further complicated by an obvious error of the temporary rescue director. There is no room for egotism in emergency cases. In any rescue, the temporary director *must* yield to a more capable leader. It could mean the difference between life and death for those trapped inside a cave.

The following rules should be memorized and taken to heart as seriously as you take the condition of your ropes and your carbide lamp.

1. Always leave word of your destination — if possible, with a caver who knows how to get to the cave. If not, try to leave a map with directions to the site. Also, leave word of the time you expect to return, and *stick to it.* The rescue team will not appreciate the time and effort it takes to locate an allegedly lost caver who is actually gallavanting nonchalantly down "just one more passage."

2. Carry the phone numbers of local rescue units, if any exist, or of the National Speleological Society, in case your party must phone for help. Don't get caught in Sam's bind: The *last* source you should call is the local postmaster or telephone operator. Emergency reports communicated to the wrong agency can cause more harm than good. Especially if you plan to venture into unexplored territory, leave the number of the NSS with someone at home. The NSS and its regional grottos keep extensive files of known caves and can be of great help in searching for possible sites. Information on local caves often can be found in unlikely places — try land owners, outdoorsmen, hunters, geological surveys, forest and park rangers, local law enforcement officers, local naturalists.

3. Carry waterproof paper and pens on your trip. Should you need to send for help, the messenger must carry all pertinent information out of the cave. The following details should be put in writing *before* the messenger leaves the cave: a) directions to the cave and cavers in as much detail as possible,   b) names, addresses and conditions of the cavers,   c) types of injury,   d) distance of the cave from the road and of the caver from the cave entrance, and e) any problems the messenger is aware of i.e., difficult passages in the cave, potential for hypothermia, special equipment that the messenger can determine is needed. When the S.O.S. is sent, the messenger can avoid the possibility of forgetting details by having them written down and transferring the written information directly to the rescue director when he reaches the cave site.

4. Plan to have *at least two* experienced cavers in your group—one to join another caver to form the two-man messenger team, the other to stay within the cave with the rest of the group. Since nobody should ever go caving alone, two messengers must always be sent. To keep the lost or injured cavers calm, an experienced caver should stay behind to keep them busy warming the injured or anyone in danger of hypothermia, draining water pools, moving obstacles (rocks, etc.), for the rescue path, heating water.

5. Be prepared to yield to more experienced cavers and rescue teams when they arrive at the scene of the

emergency. The most important factor in any cave rescue is time, and time is only wasted by useless argument, panic, confusion and lack of organization. All persons at the site of the cave should be divided into teams for communication, equipment, antihypothermia, medical care, stretcher-bearing. Those who can be of no help in these functions should be told politely, but firmly, to leave the area. Extra people only cause chaos.

For anyone but the true expert, the rescue operation is limited to transmitting the vital information listed earlier. The logistics of the rescue will be directed and handled by the authorities you call to the scene. The beginner certainly should follow the preparatory rules described above, but few newcomers to caving will be able to handle the difficult technique of extricating injured victims. Rather than describing the methods by which such difficult rescues are effected, the following will give you an idea of the basic steps involved in a typical rescue.

Cave rescue becomes necessary in two cases: 1. when the cavers become lost; 2. when a caver is injured and needs help to get out of the cave.

In the first case, the cavers themselves have little recourse but to stay put. Once a search has begun, it is easier for the search party to find the lost cavers if they stay in one place. An informed rescue team will contact all sources of cave information to locate the cave in question. Starting with the NSS and its affiliates, the rescuers can proceed to check with the sources listed earlier. As a last resort in the search for the right cave, the director can broadcast a plea for help over radio and television.

The first move of the search team is to organize several small groups who can sweep different areas. Maps of the areas they cover should be made, with details of special investigations noted (digging, underwater searches, etc.). If the first sweeps are futile, the mapped areas can be rechecked periodically and any new findings recorded. If the cavers are known to be beginners, the rescuers probably will look first in passages that could trick the uninitiated. If the search party knows that the cavers are experienced, it will probably be forced to explore areas of more difficult access, often in virgin territory. If the lost cavers are lucky, this logical, methodical search will lead rescuers to them in time. The search will go on until the party is found or the search party has lost all hope.

A cave search can require tremendous manpower, time and energy, and the rescue of an injured caver can involve the most advanced cave technique known. Extricating an immobile caver from a tight spot can be dangerous and complex. When a caver is injured, two messengers (at least one of them an experienced caver) must be dispatched, each carrying a copy of the information vital to the case. These messengers must realize that they carry the fate of those left behind in their hands and must proceed with utmost caution; their safe exit from the cave is necessary to the victim's survival. The messengers then proceed to the nearest phone, call the appropriate sources and return to the site, where the more experienced of the two takes the job of temporary rescue director. When expert help arrives, the temporary director should yield to a higher authority (especially one with a title that tells of his expertise), and do everything he can to give the rescue unit helpful information.

Chapter 8

Taking It Home on Film

Of all human frailties, perhaps memory is the most frustrating. One may recall the subtle beauties of a cavern just explored, but how long will these images remain? Naturalists may be tempted to knock off a chunk of stalactite or take home a jar of bat guano, but if everybody did that, there would be little left of the phenomena that draw thousands into caves year after year. Instead, do what Brother Eastman suggests: Record the good times on film.

At its simplest, cave photography is hardly more complicated than taking a photo in the comfort of your own living room. All you need is a camera, some film and a light source. Certain logistical problems make cave photography tricky at times but, usually, you'll be able to get your equipment in and out without serious difficulty. For beginners, all that is needed is an inexpensive Instamatic-type camera with flash cubes. Simply pull out your camera and snap away. Of course, for photo bugs and those times when you find a room that is too large for such simple equipment, more complex gear is needed. First, however, the reasons for cave photography.

Why Take Photos?

If there is one theme in caving that must be stressed, it is *conservation* Too many caves have been ruined by souvenir seekers and opportunistic businessmen. Vandalism, too, is an ever-present problem that cavers must strive to prevent. Surely, at times, the motives are innocent enough, but that naivete is born of ignorance of the need to preserve features that required millions of years to create.

To preserve the beauty and share it with even more people, the logical solution is to record nature's handiwork on film. Moreover, a small piece of rock is like a sentence out of context—nobody can truly relate it to the whole.

Pictures, on the other hand, transport viewers to that time and place in the depths of the cave. Slides, prints and movies can bring the whole experience home to your companions, and they will help you remember the time you stood on the lip of that 200-foot well of inky darkness. They may even help you recruit more caving enthusiasts.

Instamatics

For simplicity and economy, the Instamatic-type (cartridge-loading) camera is a good choice for cavers. Generally, cameras in this class cost $15 to $30, depending on how fancy they get. The advantages of the camera include ease of loading—no fumbling around with film tabs—compactness and few maintenance requirements. However, a camera of this type will have a plastic body, and plastic breaks easily, so you must transport it carefully. Of course, if you break the Instamatic, well, you are only out about 20 bucks.

Taking a cave photo with an Instamatic camera is no different than taking a shot of Aunt Maude and Uncle Josiah on the living room sofa. Just clip on a flash cube (or bar) and shoot. It's that simple, but you have to remember that a flashcube illuminates an area only up to 14 feet from the camera. Thus, if you are in a large room, you may not be able to get the shot you'd like. In addition, most Instamatics cannot focus from less than three feet. So with an Instamatic, you might as well forget about taking shots of small features that require close work. Yet, for its price and weight, the Instamatic-type camera allows great utility.

One suggestion for use of Instamatics underground: Make sure you carry your flash cubes in a shockproof container, especially if they are *Magicubes.* These cubes have a small amount of flash powder in them, and if you fall on a live cube, you are liable to end up with some nasty burns. (If you fall on your camera, you'll just get bruised.) To protect yourself, stick those cubes in a bandage tin and carry the box in your pack, not in a shirt pocket. Other than that precaution, there are no major concerns when using an Instamatic.

Instamatic films are manufactured by a variety of companies, but you would be wise to stick with the major names—Kodak, GAF, Fuji. They will give the most uniform results and can be processed by most commercial film labs. Sometimes, though, you are sure to run out of film and find yourself just outside of Nowhere, Noplace, and you will end up with an unknown brand of film. For processing these, your best bet is to send them to a reputable lab—and pray a lot.

For your initial forays into the depths, slide film is advised. Slides can be processed for a much lower cost, and you can have prints made form slides for a little extra. Prints, on the other hand, should be reserved for the most experienced photographers. Stick with a medium speed slide film—Kodak Ektachrome X, Kodachrome R, GAF 64, Fujichrome R 100. These films will give you the latitude for a broad range of colors.

35 Millimeter Photography

Talking about the merits of 35 mm photography is somewhat akin to opening Pandora's box. Everybody will tell you something different. Millions of 35s are manufactured and sold every year, and there is no one answer to any one problem. Nonetheless, here is an effort to classify just what sort of gear you need if you go in for more advanced cave photography. While the advantages of the various pieces of equipment can be debated, there is no questioning the fact that the actual situation dictates the equipment chosen.

The 35 mm format offers several advantages over the use of Instamatics. First, the ability to control speed of shutter and degree of exposure makes a 35 mm a more versatile camera for the serious cave photographer. You are able to determine just what sort of photograph you will take by judiciously manipulating the shutter speed and aperture controls. Also, the 35 mm single-lens reflex format allows you to select a variety of lens combinations and utilize different flash setups. Additionally, most good 35s are equipped with a timed exposure or bulb setting for longer than normal exposures. And, the variety of films available for 35s is almost endless—slide films, infrared, black and white, color print, high contrast and many more. The effects of the different films will not be explained here. For that you need many books, or a friendly and patient camera person. If you can find either, and are at all interested in photography, it may be worth the extra time.

The 35 mm format does, however, have certain disadvantages, cost among them. A good 35 mm camera costs about $180—quite a lot of money to drop into a pit. The market for used cameras is rather tight, but if you find a good one, you can cut your costs. Whether new or used, though, expect to spend at least $120 on the camera. Lenses are another story, and an accessory lens costs anywhere form $30 for the cheap stuff to hundreds of dollars for the better models. Face it, a solid 35 mm system will set you back a pretty penny (hundreds and hundreds of them, in fact).

A good compromise is a used camera, with Pentax, Minolta and Canon probably offering the best value in the used market today. These models provide durability and quality at a respectable and reasonable price. In general, avoid new cameras and Nikons and Leicas of any sort. Their quality is not to be questioned, but they are too expensive to take into the accident/loss-prone environment of a cave. Never take anything into a cave you wouldn't be willing to lose, and this is very true of camera equipment. The odds are good that the piece will be dunked in water, sloshed in mud or bounced like a soccer ball.

Other bits of gear required for effective cave photography include a tripod to support the camera for long-duration exposures, a cable release to minimize vibrations of opening the shutter by hand, one or more flash units and you. All these pieces combine to form the basic 35 mm caving camera.

How to Take the Shots Down There

First, you have to get the outfit into the cave. Bulky as it may seem, the gear can be taken in and out with minimal difficulty and damage. Since the camera is a delicate piece of equipment, it should be packed in a shockproof container of some sort. If a Gurnee Can is not available, a three-pound coffee can will do very well (provided you can find anyone these days able to afford coffee in this quantity). The flash unit, either bulb or electronic (electronic is recommended for versatility and safety), can be packed with the camera, and foam padding in the can will reduce the jarring effects of being dragged around the cave in your pack. Spare lenses can be included in a similar container, but one lens usually will suffice. The tripod is easy to transport once you tie the legs together to prevent them from opening and blocking the passage like an umbrella in a hallway, and the cable release can be packed with your camera or anything else you may feel deserves its company.

Now, if there is one thing the experts agree on about caves, it is that they are dark—very dark. Thus, any light needed for pictures must be provided by the caver. While the vastness of many rooms normally would prevent a person from taking a flash picture with any success, a tripod, a 35 mm camera and the other pieces of equipment previously mentioned should give you acceptable results after a few tries. Of course, the wise photographer, when embarking on a totally new set of circumstances, usually does what is called *bracketing.* Bracketing simply means taking the same shot at different exposures—one on either side of (greater or lesser than) the exposure believed to be correct. By doing this, you should be able to get one shot with acceptable lighting.

Let's take a hypothetical situation. You are in a huge room and you want to take a photo—what do you do? You know that the room is too big for your flash unit to illuminate it sufficiently, so the best idea is to set the camera on a tripod and, using the cable release, take a timed exposure longer than one second in duration. But there won't be enough light anyway, right? Wrong! First of all, you are down there with a number of cavers and they each have some source of light. However, to wait for their feeble lights to make an adequate impression on film would be analogous to waiting for Godot. Instead, take that strobe sitting on the camera and walk around the cave periodically, flashing it to give the camera enough light in the needed area. The result will be similar to those timed exposures you may have seen of lightning bolts over a city at night.

Correct exposure is crucial for color photography. While black and white film has great latitude, color can be temperamental. Correct exposure can be obtained, however, by using a little hard experience and seat-of-the-pants thinking. Your best course is to set your lens at f 2.8 or f 3.5, if you are going to use two or three flashes. Set it lower if you anticipate firing more but, as a general rule, never set the aperture below f 2.

The great thing about cave photography is that you have the light sources completely under control, so you do not have to worry about the time factor for your exposure. One caution, though: Have a friend ready to close the shutter when you are finished flashing, and make sure that his carbide is extinguished and the shutter is closed before approaching the camera. Otherwise, you will have washed-out photos.

If you are inclined toward artistic experimentation, try taking photos with the benefit of only your carbide. Generally, you should use only black and white film for efforts like these, with the lens opening set for f 2 and an exposure of about five seconds. Bracket this with exposures of four and six seconds.

Taking Moving Pictures in the Underworld (Super 8)

At times, you will want to have a more active remembrance of your trip underground. For those of you who are home-movie buffs, cave motion pictures offer an unsurpassed opportunity to expand your craft and bring home some remarkable footage.

The main problem with all movies is the need for light. At home, you can plug your favorite light into any convenient wall socket and illuminate the world. Underground, you must make do with either carbide lamps or battery powered lights.

By using a high speed black and white film (Kodak Tri-X) it is possible that if you use a wide lens opening (f2.8 or f2) you will be able to achieve adequate results with carbide or other head lamps. With the advent of automatic movie cameras (i.e., those with automatic light meters), you should not have to worry about making a series of complex settings on your camera. With an older unit, try a combination of large apperture and a slower frame speed

(12 frames per second if your camera has that adjustment).

Granted that you can get some sort of image with your headlamp, the pictures may not be up to typically acceptable standards. Perhaps the combination of low light and fast movement foils the effort. However, carbides can be extremely useful in recording the relatively slow movements of some cave life and dripping water effects.

For the more sophisticated speleocinematographer, the advent of battery-powered lights has opened new avenues, especially for color photography. No longer does the photographer have to carry massive klieg-type lights and portable gasoline generators (as the French did in the 1950s). He simply has to handle the battery pack and light with care, due to its fragility.

With proper lighting you can do virtually anything you desire—from complicated climbing sequences to an idyllic swim in an underground pool. Color will bring life to your efforts. You will be able to record the nuances of iron oxide shadings in a calcite column. The blue of cave ice will shimmer before your lens.

For use underground, experts recommend the Sylvania Rapid Charge Sun Gun #SG77. This unit is a combination flood and spot light, allowing you to illuminate a wide room or focus on a climber descending a deep well. Weighing only 3½ pounds, the unit will not place an undue stress on your shoulders (or waist). On a full charge, the light will last for 10 minutes of constant filming (that's 3 rolls of Super 8 film). It can be recharged on 110/AC or 12 VDC from your auto's cigarette lighter. There are two short-comings, however. The light itself is expensive (list $157.50), and the bulbs are expensive (type EKT, list $16.06). If you are serious about excellent cave movies, though, this is the unit to buy. But, be careful! One bump could really burn your pocketbook.

An alternative to Super 8 is 16 mm movies. These, however, are more costly, both in equipment and processing. The techniques are the same—good lightning, good composition, and proper exposure. Shooting 16 mm really should be reserved for cine students or professionals. As an amateur caver, you do not need the sophistication.

Other types of cave photography require some knowledge of the use of your equipment and of optic theory. For example, microphotography—the photography of minute features and creatures—can best be explained by a number of books and pamphlets available at your local camera store.

On the whole, cave photography can be a rewarding experience, both in the results gained and in the knowledge that you have taken something from the cave, without physically removing anything. That is the ultimate for the cave photographer. The beauty, mystery and danger are all waiting there . All that remains is for you to record it.

Chapter 9

The Joys of a Group

By now it's obvious that only certified crazies attempt solo spelunking. So where does the novice go to find other cavers, especially the kind who will spare the time to show a newcomer the ropes? The answer is your friendly neighborhood grotto, the local chapter of the National Speleological Society. Of course, local colleges and universities have outing clubs, which include caving in their activities, as do athletic clubs and Sierra Club chapters. And while most of these organizations have close relations with NSS, the core of recreational caving in the United States, it is always a better idea to get as close to the source as possible. The local grotto provides that proximity.

Grottos come in varying sizes and shapes. Some are very formal affairs that conduct exploration programs and cave surveys and publish regular reports. Others are less formal and merely serve as a way for cavers to enjoy their chosen sport together safely. Most lie somewhere in between, and with more than 100 of them nearly every area of the country has at least one grotto (see the official NSS list at the end of this chapter.) You don't have to worry about being a stranger at your first grotto meeting. If your interest in caving is genuine, one of the members at the meeting will spot you, and your fun will have just begun.

Obviously, the most important aspect of joining a grotto is that you will be in the company of experienced cavers. You will not have to go out alone, or with other relatively inexperienced cavers, thus risking life and limb. The members know the potential dangers of caving and how to compensate for a novice in their group. They also will share their experience with you. The collective skill and expertise of the grotto, as it has developed with each new member joining and learning, is there for you to draw upon. In effect, the grotto can be the "class" that will teach you, first hand, the techniques of caving and how to get the most out of the sport.

Within all this, of course, is the fellowship a grotto affords. Everyone involved in a given expedition shares the same interests and, therefore, the same enjoyment. This sharing facilitates the exchange of ideas necessary to successful cave exploration. After all, you are exploring nature, and each person views it a little differently. By sharing these views, you can only learn more about what you are doing.

On a less esoteric, but importantly fundamental plane, you will find the members of a grotto happy to share their equipment with you. Ropes, carabiners, even the clothes necessary for caving can be expensive, and the novice cannot be expected to shell out $200-$300 so he can be fully outfitted before he has even entered a cave. Granted, you cannot forever rely upon the generosity of fellow cavers, but most are understanding, realizing that caving may not be the least costly sport. They generally will let you share their equipment during your formative stages as a spelunker.

Grottos, as mentioned before, are the local chapters of the National Speleological Society and, as you get more involved in the sport of caving, you will find that NSS is where it all happens. A nonprofit organization affiliated with the American Association for the Advancement of Science, NSS was formed to advance the study, conservation, exploration and knowledge of caves. It publishes a variety of information relating to caving in this country.

In addition to the meetings held by local grottos, the society itself conducts regional conventions across the country culminating in the NSS annual convention. This event is regularly attended by nearly 600 spelunkers, and it provides the unique opportunity for an individual to meet the leading authorities of the sport and science, to discuss the latest techniques and scientific theory and to share experiences with those who are genuinely dedicated to caving.

The location of each year's convention rotates among the best possible caving regions in the country to combine meetings with field trips and explorations. Many cavers take advantage of the convention's site to camp out during the week but, for those who prefer to sleep on something softer than ground, other housing accommodations are available.

NSS also possesses one of the most complete reference libraries in the United States dealing with caving. Located at the society's Huntsville, Alabama, headquarters, the library contains some 800 volumes, and a list of circulating titles is available to all members. (Incidentally, you do not have to belong to a grotto to belong to NSS.) In addition, members have access to numerous slide programs from the Audio-Visual Aids Library, and to certain types of caving equipment.

If you plan to become a regular spelunker, you should give serious consideration to joining NSS, at least to repay the organization for what it has done for the sport. Before the society was organized in 1941, there was no concerted effort to study the vast numbers of American caves. In terms of exploring, mapping and reporting on caves, there is no single organization that has done more than NSS. You can get more information by writing NSS, Cave Avenue, Huntsville, Alabama 35810.

Regardless of whether you cave with your own group, with a grotto or other NSS affiliated venture, always keep in mind that your group is larger than it appears. A standard rule of thumb in caving is to count the members of the group and add at least one more. That extra one is the caver's constant companion—the *hodag.*

Hodags are the gremlins of the underworld and, like gremlins, descriptions of hodags vary from caver to caver. The average description of a hodag is a small, furry creature with four legs, the legs on the right side being longer than the legs on the left side. Of course, if you happen to encounter a left-handed hodag, the legs on the left side will be longer than those on the right. There is, however, one problem with all hodags. No one has ever seen one; for all practical purposes, they are invisible.

Hodags tend to visit caving expeditions at the most inopportune times. One of their favorite hiding places is high on the wall of a narrow crawlway. As each caver goes on through the crawlway, the hodag blows out the flame on the caver's carbide lamp. If you have ever tried to relight a carbide lamp in a crawlway with both arms over your head, you would know how serious it can be to be visited by these strange creatures. On the other hand, if you

have ever watched a caver try to relight a carbide lamp
with both arms over his head, you realize how much fun
hodags have. And living in the dark depths of the earth,
one can hardly fault the hodag for trying to find joy wher-
ever he can.

Other hodag antics include dropping your climbing rope
into the only muddy section to be found in the cave. The
hodag also will move the softest food you are carrying toward the bottom of your pack. There is nothing worse
than a flat, compressed tomato or wedge of cherry pie. Finally, hodags take particular glee in untying cavers' shoe-
laces, especially when the victim tries to retie his shoes
with cold, muddy hands.

The only evidence of a hodag visit, other than the obvious results, is the scurrying of little feet away from the
scene of the crime. If you move after one, it will stop and
hide, only to move again when you move, so its distance
from you is ever constant. Still, no caver really wants to
catch a hodag. They are not truly malicious creatures,
and, if one looks on the bright side, they do tend to keep
the caver on his toes.

Grottos

ALABAMA
Birmingham AL Grotto, P.O. Box 55102, Birmingham, AL 35255
Gadsden Grotto, P.O. Box 2622, E. Gadsden, AL 35903
Huntsville Grotto, Box 1702 W. Sta., Huntsville, AL 35807
Montgomery Grotto, Rt. 1, Box 24, Deatsville, AL 36022

ALASKA
Glacier Grotto, 2944 Emory St., Anchorage, AK 99508

ARIZONA
Central Arizona Grotto, 2148 Hu-Esta Dr., Tempe, AZ 85282
Escabrosa Grotto, P.O. Box 3634, Tucson, AZ 85722

ARKANSAS
Boston Mountains Grotto, Dept. of Chemistry, University of
    Arkansas, Fayetteville, AR 72701
Hendrix College Student Grotto, 1625 Clifton #3 Conway, AR 72032
M.O.L.E.S., Rt. 1, Box 14, Dogpatch, AR 72648

CALIFORNIA
Diablo Grotto, 1929 Oak Park Blvd., Pleasant Hill, CA 94523
Golden Gate Grotto, 1286 Green St., San Francisco, CA 94109
Mother Lode Grotto, Box 254545, Sacramento, CA 95865
San Diego Grotto, 8600 Lemon Ave. #9, Le Mesa, CA 92041
San Francisco Bay Chapter, Box 2282, Menlo Park, CA 94026
San Joaquin Valley Grotto, 4035 E. Dayton, Fresno, CA 93726
Santa Barbara Underground Grotto-311, 668 Willow Glen Road,
    Santa Barbara, CA 93105
Shasta Area Grotto, 131 Oleander Circle, Redding, CA 96001
Sierra Mojave Grotto, 4205 Reeves, Ridgecrest, CA 93555
Southern California Grotto, 1828 Alpha Ave., South Pasadena,
    CA 91030
Stanislaus Grotto, P.O. Box 508, Ceres, CA 95307
Waldo Brothers, 2425 Cooley Place, Pasadena, CA 91104

COLORADO
Colorado Grotto, 911 Cook St., Denver, CO 80206
Southern Col. Mountain Grotto, 13260 Ward Lane, Colorado Springs,
    CO 80908
USAFA Grotto, CWRA, USAFA, CO 80840

CONNECTICUT
Central Connecticut Grotto, 826 Church Hill Rd., Fairfield, CT 06432

DELAWARE
Cdr. Cody's Caving Club, 2603 Pecksniff Rd., Wilmington, DE 19808

FLORIDA
Florida Speleological Society, P.O. Box 12581, University Station,
    Gainesville, FL 32604
Florida State Cave Club, Box 6885, Florida State University,
    Tallahassee, FL 32306
Ft. Rucker-Ozark Grotto, 616 Mooney Rd., Ft. Walton Beach, FL 32548
Tampa Bay Area Grotto, 959 W. Cave Ct., Hernando, FL 32642

GEORGIA
Athens Speleological Society, Philoshophy Dept., Peabody Hall,
    Univ. of GA, Athens, GA 30602
Clayton Co. Cavers, 2381 Walt Stephens Rd., Jonesboro, GA 30236
Dogwood City Grotto, 1865 Ridgewood Dr., NE, Atlanta, GA 30307

IDAHO
Gem State Grotto, 2324 Pendleton, Boise, ID 83705

ILLINOIS
Rock River Speleological Society, 2225 Oxfors St., Rockford, IL 61103
Windy City Grotto, 15 South 6th Avenue, LaGrange, IL 60525

INDIANA
Bloomington IN Grotto, Swain Hall W—Physics Dept., Bloomington,
   IN 47401
Central Indiana Grotto, Box 153, Indianapolis, IN 46206
Evansville Metro. Grotto, 1025 Thompson Ave., Evansville, IN 47715
Mid-Hossier Grotto, 409 Galahad Dr., Franklin, IN 46131
Northern Indiana Grotto, 414 W. Seventh St., Auburn, IN 46706
Southern Indiana Grotto, 743 Linden Drive, Seymour, IN 47274
Speleological Studies Group, 413 S. Catherwood, Indianapolis,
   IN 46219
Western Indiana Grotto, Dept. of Geology, Indiana State University,
   Terre Haute, IN 47809

KANSAS
Kansas Speleological Society, 3462 Girard, Topeka, KS 66605

KENTUCKY
Blue Grass Grotto, Dept. of Geology, Univ. of Kentucky, Lexington,
   KY 40506
Fort Knox Grotto, 5810-B Billhymer, Fort Knox, KY 40121
Green River Grotto, Dept. Geography/Geology, Western Kentucky
   University, Bowling Green, KY 42101
Lake Cumberland Speleological Society, 116 Jonathan Lane,
   Somerset, KY 42501
Louisville Grotto, 10214 El Coco Ct., Louisville, KY 40291

LOUISIANA
Southern Mississippi Grotto, 10003 Stonehaven, Shreveport, LA 71118

MARYLAND
Annapolis Grotto, 522 Tayman Drive, Annapolis MD 21403
Baltimore Grotto, 7600 Pindell School Rd., Fulton, MD 20759
District of Columbia Grotto, 3 Pooks Hill Rd. 103, Bethesda, MD 20814
Frederick Grotto, 1715 Heather Lane, Frederick, MD 21701
Sligo Grotto, 414 N. Horners Lane, Rockville, MD 20850

MASSACHUSETTS
Boston Grotto, Box 304, Harvard Sq. Sta., Cambridge, MA 02238
New England Grotto, P.O. Box 2452, Springfield, MA 01101

MICHIGAN
Andrews University Area Grotto - 312, (Student Grotto) 610 North Main
   Street, Berrien Springs, MI 49103
Detroit Urban Grotto, 20436 Sheffield, Detroit, MI 48221
Michigan Interlakes Grotto, Box 218, Union Lake, MI 48085

MINNESOTA
Minnesota Speleological Survey, 4216 Eleventh Ave., S., Minneapolis,
   MN 55407

MISSOURI
Chouteau Grotto, 371 W. Hickam Drive, Columbia, MO 65201
Kansas City Grotto, 5086 Glenside Dr., Kansas City, MO 64129
Lake Ozarks Grotto, Lohman, MO 65053
Meramec Valley Grotto, P.O. Box 3171, University City, MO 63130
Middle MS-Valley Grotto, Box 3733, St. Louis, MO 63122
Ozark Highlands Grotto, 307 W. Portland, Springfield, MO 65807
St. Louis University Grotto, Busch Memorial Center, Rm. 301, 20 N.
  Grand Avenue, St. Louis, MO 63103
Southeast Missouri Grotto, 803 Acid Mine Rd., Sullivan, MO 63080

MONTANA
Western MT Cave Survey, P.O. Box 241, Helena, MT 59624

NEBRASKA
Greater Nebraska Grotto, 301 E. 7th, Apt. 3, Chadron, NE 69337

NEVADA
Great Basin Grotto, P.O. Box 13798, Reno, NV 89507
High Desert Grotto, 1075 Avenue D, Ely NV 89301
Southern Nevada Grotto, SR 89039, Box 53, Las Vegas, NV 89124

NEW JERSEY
Central New Jersey Grotto, 21 Kingsley Way, Freehold, NJ 07728
Northern NJ Grotto, Windy Gables, Chatham, NJ 07928

NEW MEXICO
Mesilla Valley Grotto, P.O. Box 2763, Las Cruces, NM 88004
Pecos Valley Grotto, 408 Southern Sky, Carlsbad, NM 88220
Sandia Grotto, 11700 La Cueva Ln., NE, Albuquerque, NM 87123

NEW YORK
Adirondack Grotto, 9552 Roberts Rd., R. 1, Box 441, Sauquoit,
  NY 13456
Cornell Univ. Student Grotto, Box 28, Robert Purcell Union, Cornell
  University, Ithaca, NY 14853
Helderberg Area Grotto, 2222 Campbell Avenue, Schenectady,
  NY 12306
MET Grotto, 116-32 227th St., Cambria Heights, NY 11411
Mohawk-Hudson Grotto, 1934 Fifth Avenue, Troy, NY 12180
Rensselaer Outing Club, Rensselaer Polytechnic, Troy, NY 12181
Syracuse Univ. Outing Club, Ski Lodge, Syracuse University,
  Syracuse, NY 13210

NORTH CAROLINA
Flittermouse Grotto, P.O. Box 100, Old Fort, NC 28762

OHIO
Central Ohio Grotto, 223 Fallis Rd., Columbus, OH 43214
Cleveland Grotto, 29242 Detroit Rd., Westlake, OH 44145
Dayton Area Spel. Society, 4512 Venetian Way, Dayton, OH 45439
Greater Cincinnati Grotto, 1625 Llanfair Ave., Cincinnati, OH 45224
Miami Valley Grotto, 661 Lullaby Ct., Cincinnati, OH 45238
Southern Ohio Cavers, 3330 Barnes Rd., Georgetown, OH 45121
Standing Stone Grotto, 300 N. Main, Sugar Grove, OH 43155
Wittenberg Univ. Spel. Society, Dept. of Biology, Wittenberg
    University, Springfield, OH 45501

OKLAHOMA
Central Oklahoma Grotto, 26524 Chaucer Dr., Oklahoma City,
    OK 73120
Tulsa Grotto, P.O. Box 52811, Tulsa, OK 74152

OREGON
Willamette Valley Grotto, 505 Roosevelt St., Oregon City, OR 97045

PENNSYLVANIA
Bucks County Grotto, P.O. Box 142, Warrington, PA 18976
Franklin Co. Grotto, 401 Hood St., Chambersburg, PA 17201
Greater Allentown Grotto, P.O. Box 373, Neffs, PA 18065
Huntingdon Area Spel. Society, P.O. Box 1028, Huntingdon, PA 16652
Nittany Student Grotto, Box 676, State College, PA 16801
Philadelphia Grotto, Box 2323, Mid City Sta., Philadelphia, PA 19103
Pittsburgh Grotto, P.O. Box 7565, Pittsburgh, PA 15213
Westminster College Student Grotto, Westminster College,
    New Wilmington, PA 16142
York Grotto, 625 Carbon Ave., Harrisburg, PA 17111

SOUTH DAKOTA
Paha Sapa Grotto, Dept. of Geology, SD School of Mines/Tech.,
    Rapid City, SD 57701

TENNESSEE
Chattanooga Grotto, 6647 Botsford Dr., Chattanooga, TN 37421
East Tennessee Grotto, 924 Corning Rd., Knoxville, TN 37923
Holston Valley Grotto, Box 3585 CRS, Johnson City, TN 37601
Mountain Empire Grotto, P.O. Box 842, Blountville, TN 37617
Nashville Grotto, P.O. Box 23114, Nashville, TN 37202
Smoky Mountain Grotto, Box 8297, U.T. Station, Knoxville, TN 37916
West Tennessee Chapter of NSS, 4518 Lawrence Rd., Memphis,
    TN 38122

TEXAS
Alamo Area Chapter, 14330 Oak Shadows, San Antonio, TX 78232
Bexar Grotto, 4019 Ramsgate, San Antonio, TX 78230
Dallas-Ft. Worth Grotto, P.O. Box 170274, Arlington, TX 76003
Galveston Grotto, 62 LeBrun Ct., Galveston, TX 77551
Greater Houston Grotto, 9318 WIllow Meadow, Houston, TX 77031
Permain Basin Spel. Society, 4723 W. Illinois, Midland, TX 79703
Southwest Texas Cave Club, Student Organizations Office, Southwest
    Texas State, San Marcos, TX 78666
Univ. Texas Student Grotto, Box 7672, Austin, TX 78712

UTAH
Salt Lake Grotto, 4230 Sovereign Way, Salt Lake City, UT 84117
Timpanogos Grotto, 18 E. 900 N., Provo, UT 84604
Wasatch Grotto, 4935 S. 2875 West, Roy, UT 84067

VIRGINIA
Blue Ridge Grotto, 4711 Eden Dr., N.W., Roanoke, VA 24012
Madison U. Student Grotto, Box L-38, James Madison University,
    Harrisonburg, VA 22807
Marion Area Grotto, 217 W. Chilhowie St. C-5, Marion, VA 24354
New River Valley Grotto, Dept. of Biology, Radford University,
    Radford, VA 24142
Powell Mountain Group, 914 Second Ave., E., Big Stone Gap, VA 24219
Richmond Area Spel. Society, Box 7017, Richmond, VA 23221
Shenandoah Valley Grotto, Rt. 4, Box 16, Waynesboro, VA 22980
Tidewater Grotto, P.O. Box 62642, Virginia Beach, VA 23462
Univ. Virginia Student Grotto, Box 431, Newcomb Hall,
    Charlottesville, VA 22903
VPI Grotto, Box 558, Blacksburg, VA 24060

WASHINGTON
Cascade Grotto, 1117 36th Ave., E., Seattle, WA 98112
Oregon Grotto, 13402 N.E. Clark Rd., Vancouver, WA 98665

WEST VIRGINIA
Charleston Grotto, 213 Hayes Ave., Charleton, WV 25314
Esso Grotto, 501 Ridgewood Rd., Huntington, WV 25701
Germany Valley Grotto, P.O. Box 63, Riverton, WV 26814
Greater Randolph Organization of Speleological Science,
    812 Somerset St., Lot 6, Star City, WV 26505
Greenbrier Grotto, 159 Wake Robin Tr., Lewisburg, WV 25901
Monogahela Grotto, Box 200, Barrackville, WV 26559
Mountain State Grotto, P.O. Box 246, Thomas, WV 26292
Parkersburg Area Grotto, 19 Willowbrook Acres, Parkersburg,
    WV 26101
West VA Univ. Student Grotto, Outdoor Recreation Center, Mountain
    Lair, Morgantown, WV 26506

WISCONSIN

Wisconsin Speleological Society, 1815 University Ave., Madison, WI 53706

WYOMING

Armpit Grotto, 5205 Chaparral Dr., Laramie, WY 82070
Hole-in-the-Wall Grotto, P.O. Box 1895, Mills, WY 82644
Vedauwoo Student Grotto, Box 3625 Univ. Sta., Laramie, WY 82071

FOREIGN

European Grotto, P.O. Box 9763, APO, New York, NY 09012
Toronto Grotto of NSS, Suite 802, 8 Godstone Road, Willowdale, Ontario Canada M5J 3C4

REGIONS

Arizona Regional Association, P.O. Box 42653, Tucson, AZ 85733
C.A.V.E.S. (Council of Appalachian Volunteers Engaged in Speleology), P.O. Box 842, Blountville, TN 37617
Mid-Appalachian Region, 625 Carbon Ave., Harrisburg, PA 17111
Mississippi Valley Ozark Region, 323 S. Gore Ave., Webster Groves, MO 63119
North Country Region, 3264 S. New York Ave., Milwaukee, WI 53207
Northeastern Region, 7 Garden Ave., Albany, NY 12203
Northwest Caving Association, P.O. Box 2, Elbe, WA 98330
Ohio Valley Region, Dept. of Geography, University of Kentucky, Lexington, KY 40506
Southeastern Regional Association, 5147 Bramblewood Dr., Acworth, GA 30101
Southwestern Region, 408 Southern Sky, Carlsbad, NM 88220
Texas Speleological Association, R.R. 18, Box 149-S, Auystin, TX 78726
Virginia Region, 6404 Caryhurst Dr., Oxon Hill, MD 20744
Western Region, P.O. Box 957, Shingle Springs, CA 95682

Appendix I: Glossary

ascenders,
Jumar/Gibbs—mechanical devices used to aid the caver when climbing a rope.

belay—a rope between the climber and either a fixed anchor or another caver, used to prevent the climber from injury in case he falls while climbing.

Berger Gulf—cave located in France. Holds record as the deepest cavern, more than 3,000 feet from the entrance to the deepest point of the cave.

block creep cave—slots formed by slippage between two or more massive segments of the earth's surface along a fault line.

carabiner—a metal link with a gate used to attach caver to a rope or as a braking mechanism while rapelling. Also known as snap link.

carbide lamp—enclosed light source using carbide and water to form illuminating acetylene.

chimneying—a method of ascent and descent using counterforce.

chock block—similar to piton, except it jams into existing crack spaces rather than being driven farther into the rock proper.

clastic fills—rock falls or clay deposits blocking passageways within caves.

corrasion—process of wearing away rock by windblown or water thrown material.

culupholites—round, broad, flowing shapes within a cave. Also known as draperies, shields, etc.

dissolution—the process of dissolving limestone by water to create fissures within the rock. As the water comes in contact with the limestone, the two combine to form carbonic acid which eats away at the rock.

El Sotano—cave located in Mexico. Holds the record for the longest single free drop, 1,345 feet.

gloup—passageways formed by water between the face of a cliff and the headland above it. Usually found on sea coasts. Also known as throats or blowholes.

guano—bat feces. Contains high nitrate concentrations and is, therefore, highly explosive.

helictite—formation with no defined shape. May be gypsum needles or other extremely delicate shapes.

hypothermia—condition created when the body loses more heat than it generates. One of the greatest threats to cavers.

ice caves—caverns carved out of glaciers or other formations of standing ice. They are formed by either wind or water action.

Joe Caver—a habitual caver with a haunted look when found in daylight, easily identified by a pronounced crouch as he walks. Known to flinch from overhead sounds, and has tendency to check stairs and stairwells for loose materials.

karst process—the dissolution and development of underground drainage within limestone beds. The term "karst" originated from the name of a plateau along the coast of the Adriatic Sea in Yugoslavia.

lava tubes—tubelike passageways often found in cave systems many miles in length in lava flows.

limestone cave—the most common type of all caves, found worldwide in beds of limestone laid down by prehistoric seas. The most common limestone areas in the United States are found along the Ohio and Mississippi River valleys and in the western plains states.

littoral cave—often misnamed "sea cave," and found on sea coasts rather than underwater; formed by hydraulic action or the karst process.

living cave—a cave that is still undergoing the various cave formation processes, usually involving water.

pitons—metal spikes driven into cracks in rock to provide anchor points for ropes when ascending or descending.

Prusik knot—a climbing knot used to allow the caver to ascend a standing rope.

rapelling—simplest way of descending a cave, the process of sliding down a rope using some form of braking mechanism to control the speed of the descent.

reptation—fancy name for crawling inside a cave.

sea cave—underwater cavern. Many sea caves are found off the coast of Florida and the Bahamas (the famous Blue Holes of the Carribean).

seal walking—dragging the body by pulling on upper arms and elbows.

speleogens—acute physical characteristics of caves, such as ripples (caused by flowing water) and potholes (caused by pebbles thrown into cave walls by swirling water).

speleothems—stalactites, stalagmites, helictites and culupholites.

stalactites—formations of mineral deposits, usually in the shape of icicles hanging from the ceiling created by the progressive dripping of water from a fissure in the ceiling of the cave.

stalagmites—formations "growing" from the cave floor in the form of mounds or pyramids caused by the buildup of mineral deposits left behind by dripping water.

talus caves—caves formed by blocks of rock tumbling against one another in such a fashion as to leave passageways between the individual boulders. Not usually considered to be true caves.

Appendix II: Caving Gear Check List

Don't walk out the door without first checking your equipment, not only for condition but for completeness. Following is a checklist of basic equipment. As you become more involved in caving, you may find other items to add, but these are essential:

Clothing
helmet
coveralls with pads
socks (two pairs cotton, two pairs wool)
boots
wool shirts (two)
gloves
survival kit
underwear

Personal Gear
carbide lamp
flashlight
ascenders (two)
carabiners with brake bars
canteen
first aid kit
watch
compass
pocket knife
marking pieces
ropes
slings/harnesses

Group Gear
mini-stove with sterno
cooking utensils (including can opener)
Brunton compass
shockproof thermometer
pitons, hammers, chock blocks
gurnee can

Exotica
two-way radios
inflatable boat
ice axe, ice pitons
drills
expansion bolts

INDEX